The Top 250 Planetary Nebulae

by Gary Imm

Contents

Introduction		2
What are Planetary Nebulae?		2
PNe Attributes		2
Image Acquisition		4
Image Processing		4
Image Annotation		5
Table Legend		5
The Top 250 PNe	Jan 1	6
(Transit Dates)	Mar 7	16
	Aug 8	26
	Sep 4	36
	Oct 6	46
	Dec 14	54
PNe Morphology	Spherical	56
	Bipolar	57
	Elliptical	58
	Peculiar	59
PNe Classics	Ancient	60
	Diamond Ring	61
	Owl	62
	Ring	63
	Little Dumbbell	64
PNe Features	Ansae	65
	Jets	66
	BRETs	67
	Filaments	68
	Textures	69
	End-on Torus	70
	Irregular Torus	71
	Rotating Elliptical	72
Catalogs	Messier	73
	Caldwell	74
	Minkowski	75
	Kohoutek	76
	Weinberger	77
	Abell	78
Top 250 PNe Highlights		80
Resources		81
Scholarship		81
Final Words		81
Appendix A -		
Alphabetic List of Top 250 PNe		82

Introduction

Of all of the DSO object types, planetary nebulae have the most interesting combination of structure, color and brightness. No two are alike. Their only deficient characteristic is size. Most of these objects are small and require a longer focal length setup of at least 1000 mm. A focal length setup above 2000 mm is ideal. With the right setup, these objects make excellent astrophotography targets. There are enough objects shown in this book to keep the astrophotographer busy for years.

The purpose of this book is to help the astrophotographer understand and target the best 250 PNe in the night sky.

This book provides an introduction to PNe and is primarily a visual guide to PNe. The main section contains an image, data and description for each of the top 250 PNe in the sky. Following the main section are descriptions and many posters of different features and collections of PNe.

For more detailed scientific discussions of these beautiful objects, please refer to these books - "Cosmic Butterflies" by Sun Kwok, "Planetary Nebulae" by Martin Griffiths, and "An Introduction to Planetary Nebulae" by Jason J Nishiyama.

What are Planetary Nebulae?

PNe have been observed for hundreds of years but only recently have been understood. Since even the brightest PNe cannot be seen with the unaided eye, the first PNe were not seen until the invention of the telescope. Astronomers as early as the 1700s observed these interesting objects and could determine that they were not stars. They were given the name, "Planetary Nebulae", because they looked like planets in their early telescopes.

Over time, scientists have determined that PNe are formed from gas and dust expelled from dying stars, specifically from lower-mass stars in the later red giant phase of their existence. The gas is energized by ultraviolet radiation from the hot source star. At this point, the source star is transitioning to become a white dwarf.

Based on velocity and size measurements, scientists believe that PNe have relatively short lives on the astronomical scale, something on the order of 50,000 years.

Like snowflakes, each PNe looks different. This is due to many variables, including:

- Our apparent angle of view of the PN from earth
- Mass of PN source star, which affects the driving force of the stellar winds
- Age of PN source star, which affects the chemical composition. Typically, hydrogen gas is predominant in the early phases, while in later PN life the progenitor stars are only large enough to fuse helium into oxygen and carbon.
- Structure of PN source star – single star, binary star, or multiple. Spherical PNe are believed to be the result of a single star, while more complex structures are believed to result from multi-star systems.
- Visibility of PN source star – some source stars are either too dim to see or hidden by dust and gas

The result of these variables on the PN can be grouped into 2 categories: PNe Types (broader in nature, such as bipolar or elliptical) and PNe Features (local in nature, such as jets and filaments). Some of these are addressed in the later sections of this book on PNe morphology and type.

Although scientists have learned a tremendous amount about PNe is the last 100 years, there are still many questions yet to be answered about these complex objects, as you will see in the following pages.

PNe Attributes

PNe range widely in their characteristics. A few selected comparisons are given below. Note that magnitude and distance data is unknown for about 15% of these 250 PNe, typically for the dimmest ones.

Apparent Size

The 250 PNe of this book range in apparent size from from 0.1 arc-minute in diameter (IC 2003, IC 4593, and IC 5217) to 210 arc-minutes (Sh2-210), a difference of over 2000 times. If IC 2003 was the size of the period at the end of this sentence, Sh2-210 would be 3 times as wide as this page in comparison.

The mean PNe is 1 arc-minute in diameter. For comparison, the full moon is 30 arc-minutes in diameter, and Jupiter ranges from 0.5 to 1.0 arc-minutes in diameter.

As PNe approach 0.1 arc-minute, they become bright and stellar-like, with little detail. At the other end of the spectrum, large PNe (greater than 8 arc-minutes) tend to be very faint and featureless (except for very close PNe, such as NGC 7293 - the Helix Nebula). In my opinion, the best PNe to image are generally in the 1 to 3 arc-minute range.

Actual Size

The PNe here range in actual size from from 0.14 light years in diameter (Böhm-Vitense 5-2) to 20 light years (KjPn8). PNe at the smaller end of the spectrum are young PNe, just starting to emerge. PNe at the larger end of the spectrum, greater than 5 light years in diameter, are fading while they expand. PNe beyond 8 light years in diameter are beyond the typical size limit for a PN and could be something besides a PN. There are plenty of cases in the past of large PNe which were later discovered to be an HII region or a Wolf-Rayet emission nebula.

Distance

The PNe here range in distance from 420 light years away (Böhm-Vitense 5-2) to 30,000 light years away (Abell 47). As expected, the closer PNe are generally larger and more interesting in their structure. At the distant end of the spectrum, the 15 furthest PNe all are smaller in apparent size than 0.5 arc-minutes in diameter and difficult to image well.

Visual Magnitude

The PNe here range in visual magnitude from 8.3 (brightest - NGC 7662) to 21 (dimmest - Böhm-Vitense 5-2). Since visual magnitude is a logarithmic scale, this means that the brightest PNe here is 150,000 times brighter than the dimmest!

Type and Local Features

Many aspects of PNe such as morphology, progenitors, jets, and ansae are described and shown in the sections towards the back of this book, starting on page 56.

Sky Map

The figure below shows the sky location of each of the 250 PNe of this book. As expected, for the most part the locations of these Milky Way objects in our sky align with the plane of the Milky Way from our view perspective. Note the absence of PNE overhead from February to May of each year.

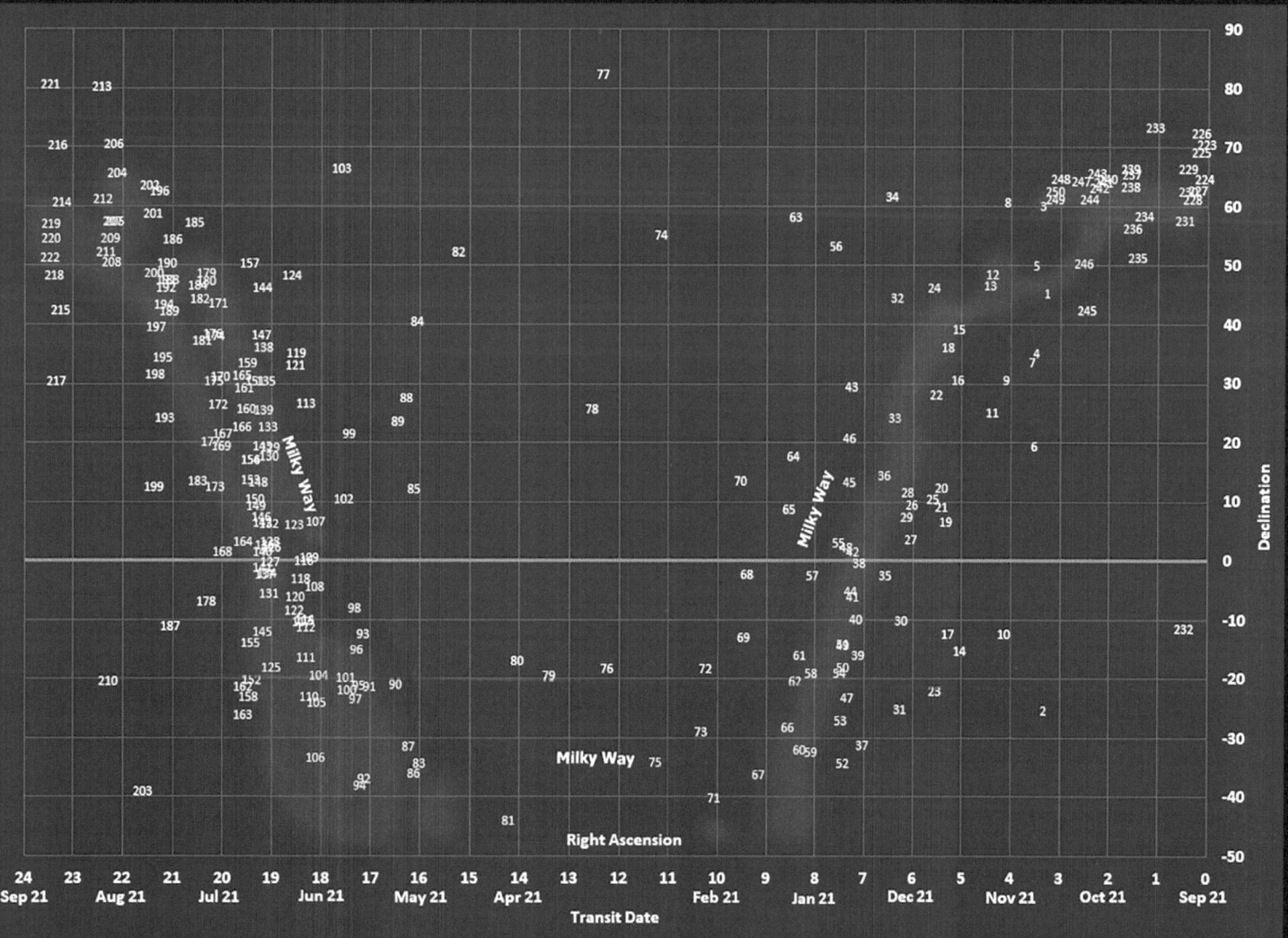

Image Acquisition

All of the PNe images of this book are shown with a north up and east left orientation to help you more effectively plan and visualize your imaging session.

All images in this book were obtained from my temporary backyard setup in Onalaska, Texas at a declination of +31 degrees and a Bortle 4/5 sky rating. The images were taken with 3 different setups:

- The majority of the images (75%) were taken with my EdgeHD 11 2800mm focal length f/10 SCT scope
- About 20% of the images were taken taken with my Tak 130 1000mm focal length f/7 refractor
- A small number (5%) of large faint PNe were imaged with my RASA 11 620mm focal length f/2.2 scope. The light gathering capability of the RASA scope was convenient for the fainter PNe.

These 3 setups represent a wide range of image scale. A medium (1000 mm) focal length scope is a nice setup for imaging most of the PNe in this book, with the exception of those below 1.5 arc-minutes in diameter. For small PNe, the 2800mm focal length is excellent but slow, making it difficult for dim objects. The fast RASA is convenient for those faint PNe that are at least 8 arc-minutes in diameter.

All cameras used for these images are ZWO CMOS mono cameras, including the ASI1600MM, ASI183MM, ASI294MM, and ASI6200MM.

With few exceptions, an total integration time of less than 5 hours was used for these images. Most of the time, the RGB channels (taken only for star color) were imaged at 2 minutes per sub and 12 subs per channel (24 minutes for each RGB channel). The OIII and HII narrowband channels were imaged at 5 minutes per sub and 12 subs per channel (1 hour total each narrowband channel).

All of the images in this book are included on my free Astrobin website, including a high resolution image and more detail on the object, the equipment used, the filters, and the exposure times.

Image Processing

Most of the images were processed using a HOO palette for the PN and a RGB palette for the stars. I used starless processing to allow aggressive stretching of the nebulae, with the stars added in later at a much lower stretch. Pixinsight, Photoshop and Lightroom were used for image processing.

A few PNe have strong enough SII signal to allow SHO narrowband processing for the PN instead of using the HOO palette, enhancing the final image with additional gold-type color tones. These PNe include Abell 21, NGC 6302, and NGC 6337. Additional PNe may benefit from this SHO approach but my normal approach to these objects used the HOO palette.

PNe which appear purely red in the following pages often contain little to no OIII content. These objects were processed using a RGB-HII palette.

Image Annotation

On each of the 250 PNe images that follows, the following parameters are displayed on each image as noted by the orange numbers:

1. Imm PNe Number (1 to 250)
2. PNe Designation
3. Distance (light years)
4. Declination (degrees)
5. Actual Width (light years)
6. Apparent Width (arc-minutes)
7. Visual Magnitude

Note that a "u" in the distance or width fields above indicates that the value is unknown.

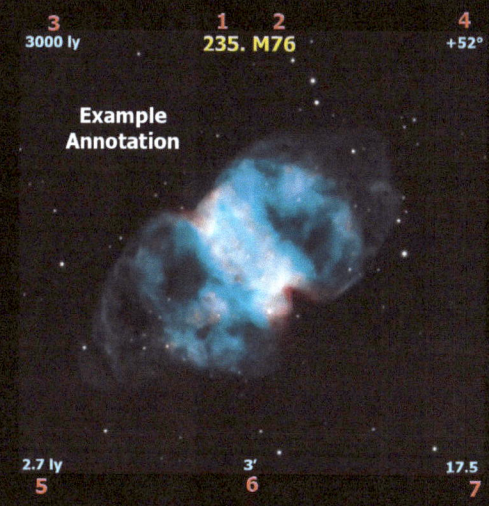

Example Annotation

Table Legend

The tables contain 14 pieces of data for each PN. A legend of the table data is provided by the orange numbers below:

Table Example:

A	B	C	D	E	F	G	H	I	J	K	L	M
#	PN Name	RA (H / M / S)	DEC (D / M / S)	Const.	Transit 9:00 PM	Transit 1:00 AM	Score	Class	Size (')	Distance (ly)	Diameter (ly)	Visual Mag
31	Abell 15	06h 27m 02s	-25° 22' 54''	CMa	Feb 16	Dec 17	2	EOY/i	0.60	12000	2.0	16.0

N *OIII dominant rim & HII dominant inner region. Rim is brighter on left than right. Bright progenitor.*

      A. Imm PNe Number (1 to 250)   F. Transit Date (see below)   K. Distance (light years, see below)
      B. PNe Name   G. Transit Date (see below)   L. Diameter (light years, see below)
      C. Right Ascension   H. Score (see below)   M. Visual Magnitude (see below)
      D. Declination   I. Class (see below)   N. Description
      E. Constellation   J. Size (arc-minutes)

Notes:

"F. Transit 9 PM" - The night of the year on which the object passes overhead at about 9 p.m., or 10 p.m. if in DST.

"G. Transit 1 AM" - The night of the year on which the object passes overhead at about 1 a.m., or 2 a.m. if in DST.

"H. Score" - ranges from 5 (showcase) to 0 (poor). The PN score is a function of the structure, size, color, detail and brightness of the object. These scores are based on those of the Compendium (see page 81). The score distribution is as follows:

      5 (Showcase - Top 2% of all sky objects)   2 (Majority of objects)
      4 (Excellent - Top 10% of all sky objects)   1 (Only for hardcore imagers)
      3 (Good - Top 25% of all sky objects)   0 (Not worth imaging)

"I. Class" - Consists of 3 capital letters indicating PN characteristics of Shape, Signal and Progenitor as follows:

    Shape: Signal: Progenitor:
    A - Ancient   O - Primarily OIII (cyan)   V - Visible
    B - Bipolar   H - Primarily Ha (red)   N - Not visible
    E - Elliptical   C - Both comparable (gray)
    M - Multipolar   R - Ha rim, OIII inner
    P - Peculiar
    S - Spherical
    X - Stellar (<0.2' diameter)

Additional PN features are listed after the "/", in lowercase as follows:

    c - bipolar lobe breakthrough   j - jets (polar)   r - ring (bright torus)
    f - bright filaments   o - owl/M97-type/inner voids   t - thin rim
    h - hexagon shape   q - opposing pair of lightened limbs on rim   y - ansae
    i - ISM interaction

"K. Distance", "L. Diameter", & "M. Visual Magnitude" - Distance and magnitude of nebulae are difficult to determine. Such values have fluctuated by 50% for some objects over the years. These values should be considered to have a wide range of uncertainty. Note that a "u" in any of these fields indicates that the value is unknown.

PNe 1-10

Jan 1

#	PN Name	RA (H / M / S)	DEC (D / M / S)	Const.	Transit 9:00 PM	Transit 1:00 AM	Score	Class	Size (')	Distance (ly)	Diameter (ly)	Visual Mag
1	HDW 03	03h 27m 15s	45° 24' 20"	Per	Jan 2	Nov 2	3	ACN	9.0	3000	8.0	17.2
	Large ancient PN, ISM affected with most visible signal in NE quadrant											
2	NGC 1360 Robins Egg	03h 33m 15s	-25° 52' 19"	For	Jan 3	Nov 3	4	POY/y	9.0	1200	3.0	11.0
	Beautifully delicate OIII PN with unique diffuse oblong appearance, small winding dust lanes and bright progenitor											
3	Kohoutek 3-94	03h 36m 08s	60° 03' 47"	Cam	Jan 4	Nov 4	1	BRN	0.20	22000	1.4	16.0
	Distant PN similar to Ring Nebula, but 10x further away and 1/10 the apparent size											
4	IC 0351	03h 47m 33s	35° 02' 48"	Per	Jan 7	Nov 7	1	XON	0.20	18000	1.0	11.9
	Tiny stellar-like PN, but some detail seen in inner region. No progenitor.											
5	IsWe 1	03h 49m 00s	50° 00' 00"	Per	Jan 7	Nov 7	2	ACN/i	20	1400	8.0	16.5
	Very faint ancient planetary nebula, distorted by ISM and discovered in 1987.											
6	Baade 1	03h 53m 37s	19° 29' 39"	Tau	Jan 8	Nov 9	2	SOY/r	1.0	8100	2.4	14.3
	Dominant OIII, progenitor star and bright mid-region rim											
7	IC 2003	03h 56m 22s	33° 52' 30"	Per	Jan 9	Nov 9	1	XON	0.1	18000	0.6	12.0
	Stellar-like OIII-dominated PN with a bit of inner detail											
8	NGC 1501 Oyster	04h 06m 59s	60° 55' 14"	Cam	Jan 12	Nov 12	4	EOY/f	1.0	5700	1.7	13.0
	OIII PN with fascinating filamentary inner structure											
9	NGC 1514 Crystal Ball	04h 09m 17s	30° 46' 33"	Tau	Jan 12	Nov 13	4	SOY/o	3.0	2000	1.9	9.5
	Impressive OIII PN with owl-type interior and bright progenitor											
10	NGC 1535 Cleopatra's Eye	04h 14m 16s	-12° 44' 22"	Eri	Jan 14	Nov 14	3	BOY	0.80	4000	1.0	12.8
	OIII bipolar PN with bright scalloped torus											

1400 ly	**5. IsWe 1**	+50°
8 ly	20'	16.5

8100 ly	**6. Baade 1**	+19°
2.4 ly	1'	14.3

18,000 ly	**7. IC 2003**	+34°
0.6 ly	0.1'	12.0

5700 ly	**8. NGC 1501**	+61°
1.7 ly	1'	13.0

2000 ly	**9. NGC 1514**	+31°
1.9 ly	3'	9.5

4000 ly	**10. NGC 1535**	−13°
1.0 ly	0.8'	12.8

PNe 11-20

Jan 20

#	PN Name	RA (H/M/S)	DEC (D/M/S)	Const.	Transit 9:00 PM	Transit 1:00 AM	Score	Class	Size (')	Distance (ly)	Diameter (ly)	Visual Mag
11	Haro 3-29	04h 37m 24s	25° 02' 44"	Tau	Jan 20	Nov 20	2	PCY	0.30	15000	1.4	15.9
	Unusual object, likely a PN but asymmetric.											
12	G156.4+01.1	04h 38m 25s	48° 38' 52"	Per	Jan 20	Nov 20	1	PHN/i	2	u	u	u
	Dim asymmetric HII-dominated PN with no progentor, likely impacted by the ISM											
13	Sh2-216	04h 43m 21s	46° 42' 06"	Per	Jan 21	Nov 21	3	ARN/i	120	420	15	12.9
	Largest PN and closest to earth. Very faint.											
14	Abell 07	05h 03m 08s	-15° 36' 13"	Lep	Jan 26	Nov 26	2	AOY/aq	13	1800	6.7	16.0
	Large faint PN, mostly OIII, with blue central star and HII ansae.											
15	Abell 08	05h 06m 38s	39° 08' 08"	Aur	Jan 27	Nov 27	1	SCN	1.0	5200	1.5	18.0
	Part of Bica 6 open cluster. Very faint with no central star. No inner structure.											
16	Kohoutek 2-01	05h 07m 16s	30° 48' 00"	Aur	Jan 27	Nov 27	2	EOY/i	3.0	3500	3.0	u
	OIII-dominated irregularly shaped PN with progenitor. Despite its PN characteristics, there is debate whether this is actually a PN.											
17	IC 0418 *Spirograph*	05h 27m 28s	-12° 41' 50"	Lep	Feb 1	Dec 2	1	ERY	0.20	5000	0.30	9.3
	Tiny stellar-like bright PN, progenitor visible											
18	Abell 09	05h 29m 00s	36° 02' 00"	Aur	Feb 2	Dec 3	2	SHN	0.50	13000	2.0	19.0
	Outer rim strong in HII. Slight dimming at 12 o'clock.											
19	Abell 10	05h 31m 48s	06° 56' 09"	Ori	Feb 2	Dec 3	2	SRY/h	0.50	13000	2.0	14.0
	Scalloped HII outer rim surrounding brighter OIII inner region. Blue progenitor star faintly visible.											
20	Haro 3-75	05h 40m 42s	12° 21' 00"	Ori	Feb 5	Dec 6	2	ERY	0.50	10000	1.5	14.0
	Bright progenitor, typical Ha rim / OIII inner signal											

11. Haro 3-29 — 15000 ly, +25°, 8 ly, 0.3', 15.9

12. G156.4+1.1 — u, +49°, u, 2', u

13. Sh2-216 — 420 ly, +47°, 15 ly, 120', 12.9

14. Abell 7 — 1800 ly, -16°, 7 ly, 13', 16.0

5200 ly	**15. Abell 8**	+39°	3500 ly	**16. Kohoutek 2-1**	+31°
1.5 ly	1.0'	18.0	3 ly	3'	u
5000 ly	**17. IC 418**	-13°	13000 ly	**18. Abell 9**	+36°
0.3 ly	0.2'	9.3	2.0 ly	0.5'	19.0
13,000 ly	**19. Abell 10**	+7°	10,000 ly	**20. Haro 3-75**	+12°
2.0 ly	0.5'	14.0	1.5 ly	0.5'	14.0

PNe 21-30

Feb 5

#	PN Name	RA (H/M/S)	DEC (D/M/S)	Const.	Transit 9:00 PM	Transit 1:00 AM	Score	Class	Size (')	Distance (ly)	Diameter (ly)	Visual Mag
21	NGC 2022	05h 42m 06s	09° 05' 13"	Ori	Feb 5	Dec 6	2	BOY	0.50	8000	1.1	14.2
	Kissing Crescents — Bipolar OIII PN with scalloped torus											
22	Pu 1	05h 52m 48s	28° 06' 00"	Tau	Feb 8	Dec 9	1	EHN	1.0	6800	2.0	18.0
	Very faint HII-dominated elliptical PN											
23	DeHt 1	05h 55m 00s	-22° 30' 00"	Lep	Feb 8	Dec 9	2	SOY/it	2.0	6200	3.5	u
	Delicate crescent-moon shaped PN. Progenitor visible.											
24	IC 2149	05h 56m 24s	46° 06' 17"	Aur	Feb 9	Dec 10	2	BOY/y	0.25	3600	0.28	11.4
	Tiny bi-polar OIII-dominated PN with red ansae and visible progenitor											
25	WeDe 1	05h 59m 00s	10° 42' 00"	Ori	Feb 9	Dec 10	2	AHY	15	1900	8.0	17.2
	Large, very faint ancient HII-dominated PN											
26	Abell 12	06h 02m 23s	09° 39' 03"	Ori	Feb 10	Dec 11	2	SRN/i	0.70	7000	1.4	12.0
	Strong HII rim. Breakthrough at 12 o'clock. Hidden in bright star glare											
27	Abell 13	06h 04m 47s	03° 56' 27"	Ori	Feb 11	Dec 12	3	EHY/ir	3.0	4000	3.5	19.9
	Strong HII rim. Little OIII signal. ISM likely interacting with nebula and blurring edges. Faint blue progenitor star.											
28	Abell 14	06h 11m 09s	11° 46' 47"	Ori	Feb 12	Dec 13	2	BHY/q	0.70	11000	2.1	15.0
	Bipolar PN. Viewing torus from near side-on view. Bright blue central star with unseen progenitor companion.											
29	Weinberger 1-04	06h 14m 34s	07° 34' 30"	Ori	Feb 13	Dec 14	2	BHN/c	0.70	15000	3.0	u
	Bipolar HII PN with lobes blown out											
30	HDW 05	06h 23m 36s	-10° 13' 00"	Mon	Feb 15	Dec 17	2	MHN/i	1.5	3600	1.5	15.4
	Looks multipolar, dominant HII composition. Faint wide HII stream reaching up. No progenitor.											

21. NGC 2022 — 8000 ly, +9°, 1.1 ly, 0.5', 14.2

22. Pu 1 — 6800 ly, +28°, 2.0 ly, 1', 18

22. DeHt 1 — 6200 ly, -21°, 3.5 ly, 2', u

24. IC 2149 — 3600 ly, +46°, 0.3 ly, 0.25', 11.4

| 1900 ly | **25. WeDe 1** | +11° | 7000 ly | **26. Abell 12** | +10° |
| 8.0 ly | 15′ | 17.2 | 1.4 ly | 0.7′ | 12.0 |

| 4000 ly | **27. Abell 13** | +4° | 11,000 ly | **28. Abell 14** | +12° |
| 3.5 ly | 3′ | 19.9 | 2.1 ly | 0.7′ | 15.0 |

| 15,000 ly | **29. Weinberger 1-4** | +8° | 3600 ly | **30. HDW 5** | −10° |
| 3.0 ly | 0.7′ | u | 1.5 ly | 1.5′ | 15.4 |

PNe 31-40

Feb 16

#	PN Name	RA (H/M/S)	DEC (D/M/S)	Const.	Transit 9:00 PM	Transit 1:00 AM	Score	Class	Size (')	Distance (ly)	Diameter (ly)	Visual Mag
31	Abell 15	06h 27m 02s	-25° 22' 54"	CMa	Feb 16	Dec 17	2	EOY/i	0.60	12000	2.0	16.0
	OIII dominant rim & HII dominant inner region. Rim is brighter on left than right. Bright progenitor.											
32	NGC 2242	06h 34m 07s	44° 46' 38"	Aur	Feb 18	Dec 19	1	BOY/o	0.50	9600	1.5	15.0
	OIII-dominated PN with bright-rimmed inner torus											
33	Minkowski 1-07	06h 37m 21s	24° 00' 36"	Gem	Feb 19	Dec 20	2	BRN	0.50	20000	3.0	13.5
	Small bipolar PN with odd OIII glow											
34	Abell 16	06h 43m 55s	61° 47' 25"	Lyn	Feb 21	Dec 22	1	SCY/i	2.4	4400	3.0	18.7
	Very faint. The progenitor star is visible as the tiny blue star just left of center.											
35	Abell 18	06h 56m 14s	-02° 53' 08"	Mon	Feb 24	Dec 25	2	ERN/hr	1.4	5000	2.0	20.9
	Outer rim stronger in HII while inner stronger in OIII. Elliptical PN, pinched at the 5 and 11 o'clock. No progenitor.											
36	Abell 19	06h 59m 57s	14° 36' 47"	Gem	Feb 25	Dec 26	2	ERY/q	1.4	7500	3.0	16.0
	Strong HII rim, brightest on left and right sides. Central star is likely in foreground and not the progenitor.											
37	Minkowski 3-01	07h 02m 50s	-31° 35' 30"	CMa	Feb 25	Dec 27	3	BON/y	0.50	15000	2.2	13.0
	Bipolar PN with ansae from the jet outflows											
38	NGC 2346	07h 09m 22s	-00° 48' 22"	Mon	Feb 27	Dec 28	3	BRY/c	2.0	4800	3.0	11.6
	Butterfly — Complex bipolar PN with HII rim and OIII center, progenitor is visible											
39	Kohoutek 1-10	07h 12m 36s	-16° 06' 00"	CMa	Feb 28	Dec 29	2	BHN	1.5	9000	4.0	u
	Faint HII-dominated bipolar PN with no progenitor											
40	Weinberger 1-06	07h 17m 24s	-10° 07' 00"	Mon	Mar 1	Dec 30	1	EHY	1.5	4600	2.0	u
	Elliptical HII PN with progenitor visible											

31. Abell 15 — 12,000 ly — -25° — 2.0 ly — 0.6' — 16.0

32. NGC 2242 — 9600 ly — +45° — 1.5 — 0.5' — 15.0

33. Minkowski 1-7 — 20,000 ly — +24° — 3 ly — 0.5' — 13.5

34. Abell 16 — 4400 ly — +62° — 3 ly — 2.4' — 18.7

5000 ly	**35. Abell 18**	-3°	7500 ly	**36. Abell 19**	+15°
2.0 ly	1.4'	20.9	3 ly	1.4'	16.0
15,000 ly	**37. Minkowski 3-1**	-32°	4800 ly	**38. NGC 2346**	-1°
2.2 ly	0.5'	13.0	3.0 ly	2'	11.6
9000 ly	**39. Kohoutek 1-10**	-16°	4600 ly	**40. Weinburger 1-6**	-10°
4.0 ly	1.5'	u	2.0 ly	1.5'	u

PNe 41-50

Mar 2

#	PN Name	RA (H/M/S)	DEC (D/M/S)	Const.	Transit 9:00 PM	Transit 1:00 AM	Score	Class	Size (')	Distance (ly)	Diameter (ly)	Visual Mag
41	PFP 1	07h 22m 00s	-06° 18' 00"	Mon	Mar 2	Dec 31	1	ARY	25	1700	12	u
	Huge ancient PN, believed to be 70,000 years old											
42	Abell 20	07h 22m 58s	01° 45' 37"	CMi	Mar 2	Jan 1	1	SOY/r	1.0	6500	2.0	16.4
	Almost perfectly circular, with strong OIII rim. Progenitor star is centered exactly and beaming bright blue.											
43	NGC 2371	07h 25m 34s	29° 29' 17"	Gem	Mar 3	Jan 1	4	BOY/cy	2.0	5000	3.2	13.5
	Double Bubble	Bipolar OIII PN with blown out lobes and a visible progenitor.										
44	Minkowski 3-03	07h 26m 34s	-05° 21' 52"	Mon	Mar 3	Jan 2	2	BRN	0.30	19000	1.6	u
	Distant small bipolar PN with HII-dominated edges											
45	Abell 21	07h 29m 03s	13° 14' 48"	Gem	Mar 4	Jan 2	5	ARN/if	8.0	1500	3.5	16.0
	Medusa	Asymmetric, likely due to ISM. Faint arcs lower right mirror the brighter arcs at upper left.										
46	NGC 2392	07h 29m 11s	20° 54' 42"	Gem	Mar 4	Jan 2	4	BHY/f	1.0	6000	1.7	9.7
	Eskimo	Tremendous detail throughout this bipolar PN with balanced OIII and HII signals.										
47	KW 8	07h 33m 25s	-23° 26' 09"	Pup	Mar 5	Jan 3	1	EHY/i	2.0	u	u	u
	Faint HII-dominated elliptical PN with ISM interaction											
48	Abell 22	07h 36m 06s	02° 24' 00"	CMi	Mar 6	Jan 4	3	BHN/c	2.0	2800	1.6	19.6
	Side-on view of bi-polar PN. Bright central torus is only part of nebula that can be seen visually. No progenitor.											
49	NGC 2438	07h 41m 51s	-14° 44' 05"	Pup	Mar 7	Jan 5	3	ERY	1.5	4200	1.9	11.5
	PN near bright cluster M46											
50	NGC 2440	07h 41m 55s	-18° 12' 31"	Pup	Mar 7	Jan 5	3	MRN	1.3	4000	1.6	10.1
	Albino Butterfly	Complex multipolar PN with HII rim and OII center. No progenitor										

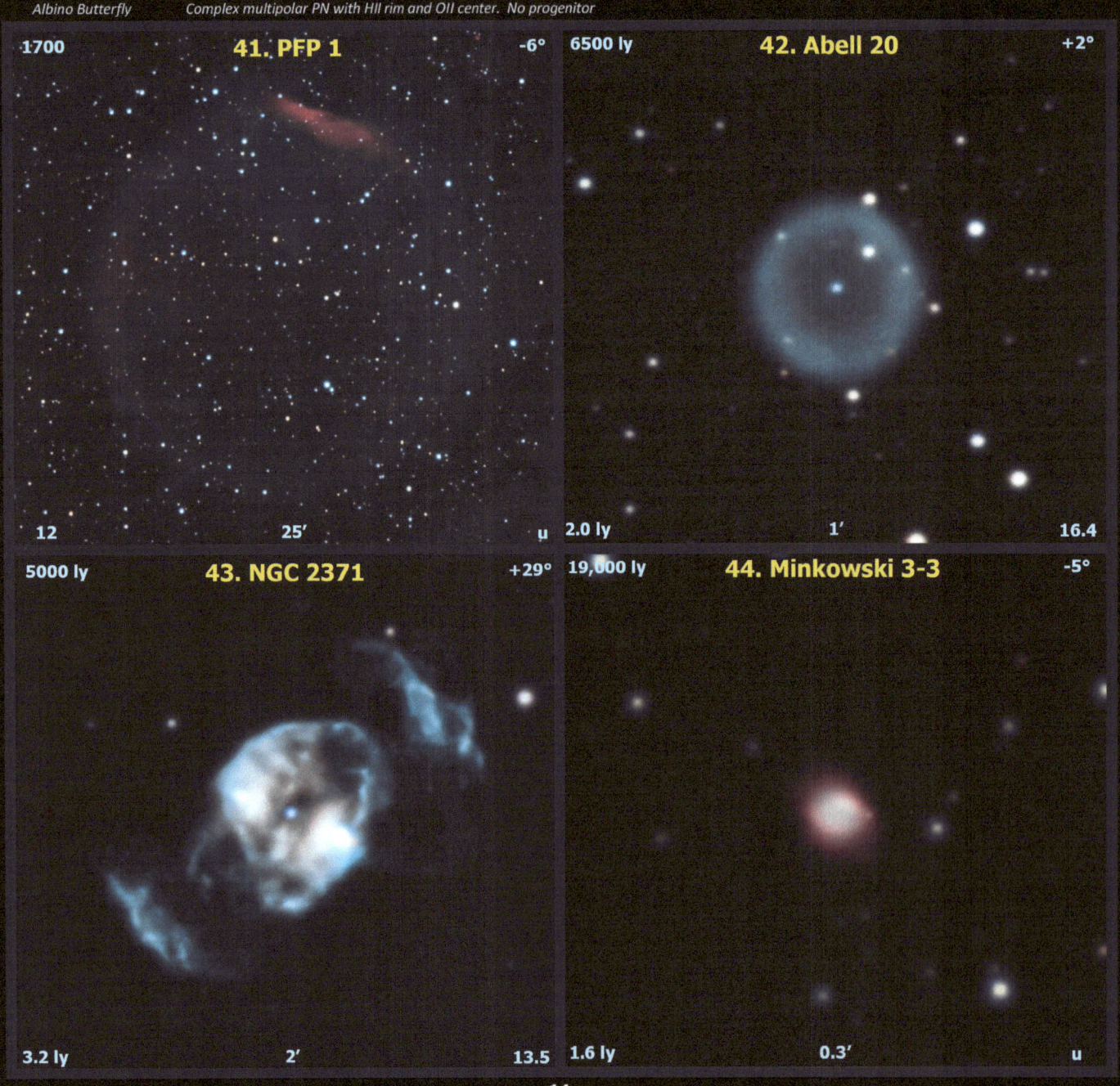

1500 ly	**45. Abell 21**	+13°
3.5 ly	8'	16.0

6000 ly	**46. NGC 2392**	+21°
1.7 ly	1'	9.7

u	**47. KW 8**	-23°
u	2'	u

2800 ly	**48. Abell 22**	+2°
1.6 ly	2'	19.6

4200 ly	**49. NGC 2438**	-15°
1.9 ly	1.5'	11.5

4000 ly	**50. NGC 2440**	-18°
1.6 ly	1.3'	10.1

PNe 51-60

Mar 7

#	PN Name	RA (H/M/S)	DEC (D/M/S)	Const.	Transit 9:00 PM	Transit 1:00 AM	Score	Class	Size (')	Distance (ly)	Diameter (ly)	Visual Mag
51	Minkowski 1-18	07h 42m 06s	-14° 12' 00"	Pup	Mar 7	Jan 6	2	ERN/ir	0.50	15000	2.2	14.0
	Small PN with interesting hole punch look and red HII rim											
52	Abell 23	07h 43m 19s	-34° 45' 13"	Pup	Mar 8	Jan 6	1	SRN/ho	1.0	7500	2.2	u
	Very faint. Circular and primarily OIII signal. Rim is slightly stronger in Ha. Little structure is apparent.											
53	NGC 2452	07h 47m 26s	-27° 20' 07"	Pup	Mar 9	Jan 7	2	PON	0.50	11400	1.7	12.0
	Peculiar PN, with bright white "S" in center of OIII-dominated PN											
54	Kohoutek 1-12	07h 50m 12s	-19° 18' 16"	Pup	Mar 9	Jan 8	2	ERN/q	1.0	9000	2.7	16.0
	Elliptical PN with HII rim and no progenitor											
55	Abell 24	07h 51m 38s	03° 00' 27"	CMi	Mar 10	Jan 8	3	AHY/q	7.0	2300	4.5	17.2
	HII-dominated. Looked more like HII region than PN. Progenitor star appears to be the small blue star at center.											
56	Jones-Emberson 1 *Headphones*	07h 57m 54s	53° 25' 00"	Lyn	Mar 11	Jan 10	4	ERY/qr	6.0	1600	3.0	17.0
	HIII-dominated ring-shaped elliptical PN with progenitor and nice texture											
57	Abell 25	08h 06m 45s	-02° 52' 43"	Mon	Mar 14	Jan 12	2	EOY/q	3.0	2500	2.2	18.0
	PN is OIII-dominant & elliptical in shape. Strong signal in the 2 opposite rim segments. Progenitor star visible.											
58	Sa 2-21	08h 08m 42s	-19° 08' 00"	Pup	Mar 14	Jan 12	2	ERN/o	0.90	11000	3.0	13.7
	Elliptical PN very similar to the Owl Nebula, M97											
59	Abell 26	08h 09m 01s	-32° 40' 15"	Pup	Mar 14	Jan 12	2	SRN/hr	0.70	11000	2.1	18.0
	Hint of hexagon shape to rim. No progenitor star. Like many PN, the outer rim is stronger in HII.											
60	Abell 27	08h 31m 53s	-32° 06' 07"	Pyx	Mar 20	Jan 18	2	EHN/hq	1.0	7700	2.1	16.0
	HII-dominant. No progenitor. 2 pairs of opposing bright regions. Could be bi-polar PN with 2 outflow axes.											

51. Minkowski 1-18 — 15,000 ly, -14°, 2.2, 0.5', 14.0

52. Abell 23 — 7500 ly, -35°, 2.2 ly, 1', u

53. NGC 2452 — 11,400 ly, -27°, 1.7 ly, 0.5', 12.0

54. Kohoutek 1-12 — 9000 ly, -19°, 2.7 ly, 1.0', 16.0

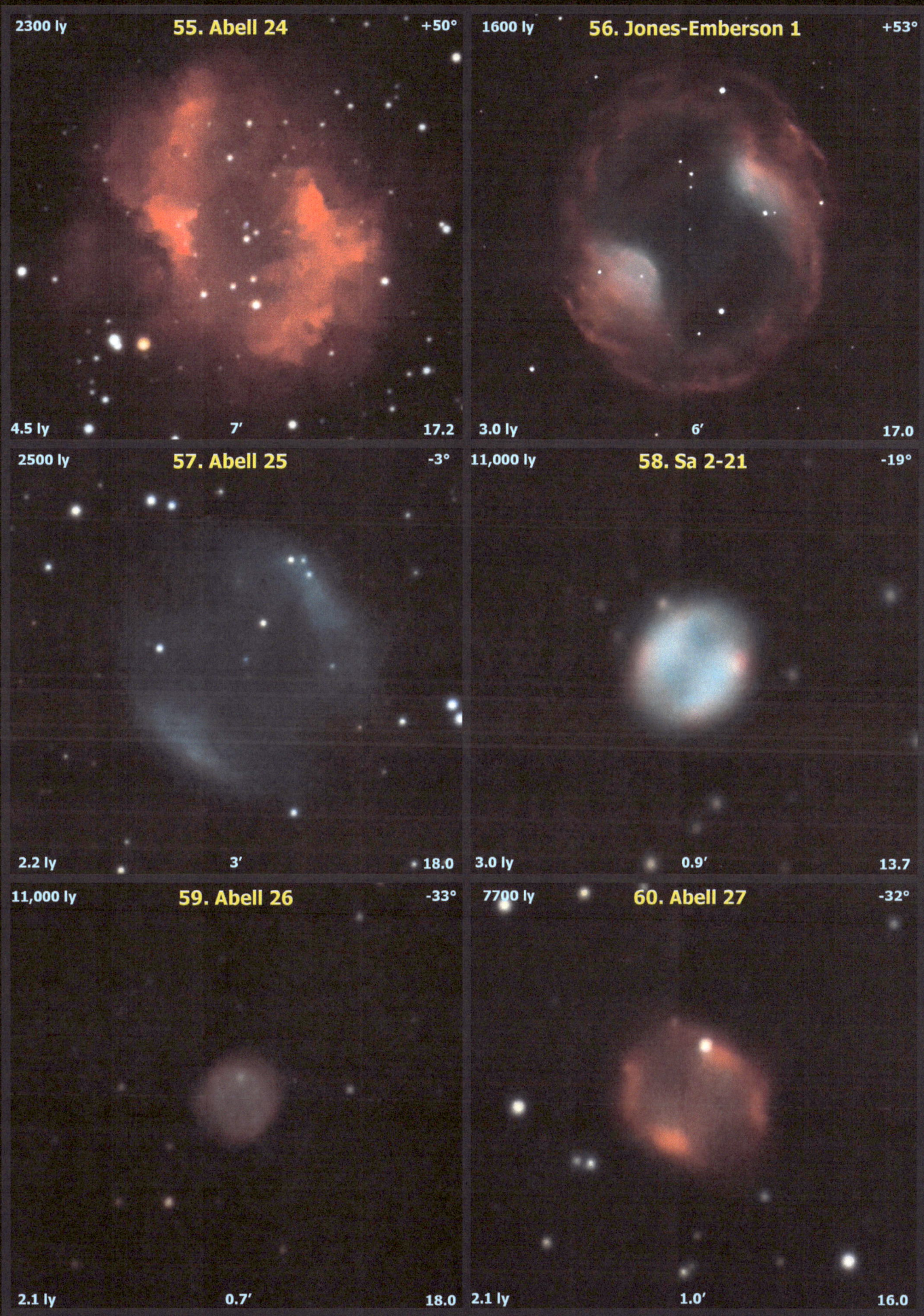

Mar 20

PNe 61-70

#	PN Name	RA (H / M / S)	DEC (D / M / S)	Const.	Transit 9:00 PM	Transit 1:00 AM	Score	Class	Size (')	Distance (ly)	Diameter (ly)	Visual Mag
61	NGC 2610	08h 33m 23s	-16° 08' 57"	Hya	Mar 20	Jan 19	2	BOY/r	1.2	7000	2.5	12.7
	Small OIII PN offers significant detail, including unusual inner HII signal											
62	Abell 29	08h 40m 14s	-20° 53' 41"	Pyx	Mar 22	Jan 20	2	ARY/q	6.0	3600	6.0	18.3
	Outer HII rim. Inner section stronger in OIII. 18.2 mag blue central star. Bright regions at 4 & 10 o'clock.											
63	Abell 28	08h 41m 35s	58° 13' 54"	UMa	Mar 22	Jan 21	1	AHY/i	8.0	1250	3.0	16.6
	Extremely dim and large. HII-dominant. Brilliant blue progenitor. Concentric set of arcs at lower right.											
64	Abell 30	08h 46m 54s	17° 52' 33"	Cnc	Mar 24	Jan 22	3	SOY/f	2.0	7000	4.0	14.3
	Central star easily seen. Interesting OIII emissions seen in inner region, forming pinwheel-type patterns.											
65	Abell 31	08h 54m 13s	08° 53' 59"	Cnc	Mar 26	Jan 24	4	ARY/o	15	1800	8.0	15.5
	Largest Abell PN with blue central star and HII rim.											
66	Kohoutek 1-02	08h 57m 48s	-28° 36' 00"	Pyx	Mar 27	Jan 25	3	POY/jy	1.8	9000	5.0	15.3
	OIII-dominated PN with rare visible bi-polar jet from progenitor											
67	NGC 2818	09h 16m 01s	-36° 37' 37"	Pyx	Mar 31	Jan 29	4	BRN	2.0	10000	6.0	11.2
	Bipolar nebula with HII rim and OIII interior											
68	Abell 33 Diamond Ring	09h 39m 09s	-02° 48' 33"	Hya	Apr 6	Feb 4	4	SOY/ot	4.0	2500	3.0	13.0
	Foreground rim star is 7.2 mag star HD 83535. Progenitor star, just above and right of center, is a double star.											
69	Abell 34	09h 45m 35s	-13° 10' 14"	Hya	Apr 8	Feb 6	2	SOY/qt	5.0	2400	3.3	16.3
	OIII-dominated. 2 limbs seen at 4 and 11 o'clock. Tiny galaxy seen in rim at 3 o'clock. Central star visible.											
70	EGB 06	09h 53m 00s	13° 45' 00"	Leo	Apr 10	Feb 8	1	ARY/iq	12	2100	8.0	16.0
	Faint ancient PN. One of the largest PN at 8 light years diameter.											

61. NGC 2610 — 7000 ly, -16°, 2.5 ly, 1.2', 12.7

62. Abell 29 — 3600 ly, -21°, 6 ly, 6', 18.3

63. Abell 28 — 1250 ly, +58°, 3 ly, 8', 16.6

64. Abell 30 — 7000 ly, +18°, 4 ly, 2.0', 14.3

| 1800 ly | **65. Abell 31** | +9° |
| 8 ly | 15' | 15.5 |

| 9000 ly | **66. Kohoutek 1-2** | −29° |
| 5 ly | 1.8' | 15.3 |

| 10,000 ly | **67. NGC 2818** | −37° |
| 6 ly | 2' | 11.2 |

| 2500 ly | **68. Abell 33** | −3° |
| 3 ly | 4' | 13.0 |

| 2400 ly | **69. Abell 34** | −13° |
| 3.3 ly | 5' | 16.3 |

| 2100 ly | **70. EGB 6** | +14° |
| 8 ly | 12' | 16.0 |

Apr 13

PNe 71-80

#	PN Name	RA (H/M/S)	DEC (D/M/S)	Const.	Transit 9:00 PM	Transit 1:00 AM	Score	Class	Size (')	Distance (ly)	Diameter (ly)	Visual Mag
71	NGC 3132	10h 07m 02s	-40° 26' 11"	Vel	Apr 13	Feb 11	4	MRY/r	1.2	2800	1.0	10.0
	Eight-burst											
72	NGC 3242	10h 24m 46s	-18° 38' 34"	Hya	Apr 18	Feb 16	4	EOY/r	1.0	4000	1.1	12.1
	Ghost of Jupiter											
73	Kohoutek 1-28	10h 34m 30s	-29° 07' 00"	Hya	Apr 20	Feb 18	1	SOY	1.0	6600	2.0	16.0
74	M 097	11h 14m 48s	55° 01' 08"	UMa	Apr 30	Feb 28	5	BRY/o	3.8	2000	1.8	15.8
	Owl											
75	Kohoutek 1-22	11h 26m 42s	-34° 22' 00"	Hya	May 3	Mar 3	3	SOY/ot	3.0	3000	2.8	16.1
76	NGC 4361	12h 24m 31s	-18° 47' 06"	Crv	May 18	Mar 18	3	BOY/j	1.8	1250	0.70	13.2
77	IC 3568	12h 33m 07s	82° 33' 49"	Cam	May 20	Mar 20	1	EOY	0.20	12000	0.70	10.6
	Lemon Slice											
78	LoTr 5	12h 56m 00s	25° 53' 00"	Com	May 26	Mar 26	2	AOY/o	9.0	1600	4.2	8.9
79	Abell 36	13h 40m 41s	-19° 52' 57"	Vir	Jun 6	Apr 6	2	ACY/f	8.0	800	2.0	12.0
80	Abell 37	14h 04m 26s	-17° 13' 41"	Vir	Jun 12	Apr 12	2	ERY/h	1.0	8000	2.3	13.9

Row notes:
- 71: One of the most spectacular PN in the sky, nicknamed the Southern Ring and the Eight-Burst Nebula
- 72: Bright OIII PN, likely bipolar, with visible progenitor
- 73: Faint PN with progenitor. Bright rim arc at 11 o'clock
- 74: Showcase PN, big and bright with inner void areas and HII rim accents
- 75: Lovely OIII-dominant PN with thin rim and inner texture
- 76: Small OIII PN with polar jets and visible progenitor
- 77: Tiny OIII-dominated PN with progenitor and inner shell
- 78: Owl-shaped OIII-dominated PN with bright progenitor that looks like foreground star
- 79: The closest Abell PN. Large, very faint, odd non-PN like structure. OIII-dominated in outer shell.
- 80: HII rim and OIII interior. Nice structure for small object. Upper left quadrant diffuse. Faint central star.

71. NGC 3132 — 2800 ly, -40°, 1.0 ly, 1.2', 10.0

72. NGC 3242 — 4000 ly, -19°, 1.1 ly, 1', 12.1

73. Kohoutek 1-28 — 6600 ly, -29°, 2 ly, 1', 16.0

74. M97 — 2000 ly, +55°, 1.8 ly, 3.8', 15.8

3000 ly	**75. Kohoutek 1-22**	-34°	
2.8 ly	3'	16.1	
1250 ly	**76. NGC 4361**	-19°	
0.7 ly	1.8'	13.2	
12,000 ly	**77. IC 3568**	+83°	
0.7 ly	0.2'	10.6	
1600 ly	**78. LoTr 5**	+26°	
4.2 ly	9'	8.9	
800 ly	**79. Abell 36**	-20°	
2.0 ly	8'	12.0	
8000 ly	**80. Abell 37**	-17°	
2.3 ly	1'	13.9	

Jun 17

PNe 81-90

#	PN Name	RA (H / M / S)	DEC (D / M / S)	Const.	Transit 9:00 PM	Transit 1:00 AM	Score	Class	Size (')	Distance (ly)	Diameter (ly)	Visual Mag
81	IC 4406	14h 22m 26s	-44° 09' 04"	Lup	Jun 17	Apr 17	4	BRN	1.8	9000	5.0	11.0
	Box	Side-on view of bipolar PN, no progenitor, no lobe breakthrough										
82	Jacoby 1	15h 22m 00s	52° 22' 00"	Boo	Jul 2	May 2	1	AOY	10	2600	7.0	16.6
		Very faint OIII-dominated ancient PN with progenitor visible										
83	NGC 6026	16h 01m 21s	-34° 32' 37"	Lup	Jul 12	May 12	2	EOY/hr	1.0	9500	3.0	12.5
		OIII PN with diffuse hexagon rim, hint of an inner ring, and some structure										
84	NGC 6058	16h 04m 27s	40° 40' 59"	Her	Jul 13	May 13	2	MOY/r	0.70	11500	2.2	12.9
		Multipolar OIII PN with progenitor										
85	IC 4593	16h 11m 44s	12° 04' 19"	Her	Jul 15	May 15	1	XOY	0.10	10500	0.30	10.7
	White Eyed Pea	Stellar-like blue PN with progenitor visible										
86	NGC 6072	16h 12m 58s	-36° 13' 47"	Sco	Jul 15	May 15	2	BRY	1.8	4000	2.0	12.1
		Small bipolar PN with HII rim, OIII central region, and progenitor visible										
87	Abell 38	16h 23m 17s	-31° 44' 57"	Sco	Jul 17	May 17	3	EHN/f	2.5	2000	1.5	20.0
		HII-dominated with very little OIII present. No progenitor. HII elliptical ring distorted, likely by interaction with ISM.										
88	Abell 39	16h 27m 33s	27° 54' 34"	Her	Jul 19	May 19	4	SOY/ot	3.0	4000	3.5	15.6
		OIII-dominated. Central star visible. Nebula has thin rim and texture. Galaxy at 2 o'clock inside nebula, 1.7 bly away.										
89	NGC 6210	16h 44m 30s	23° 47' 59"	Her	Jul 23	May 23	3	BOY/fjy	0.80	6500	1.4	11.7
	Turtle	Bipolar OIII PN with jets, ansae and filaments										
90	Abell 40	16h 48m 34s	-21° 00' 40"	Oph	Jul 24	May 24	2	EOY/r	0.60	13000	2.2	18.0
		Progenitor star visible. Both OIII and HII present, but OIII much stronger. Elliptical nebula has thick outer rim.										

9000   81. IC 4406   -44°
5 ly   1.8'   11.0

2600 ly   82. Jacoby 1   +52°
7 ly   10'   16.6

9500 ly   83. NGC 6026   -35°
3 ly   1'   12.5

11,500 ly   84. NGC 6058   +41°
2.2 ly   0.7'   12.9

10,500 ly	**85. IC 4593**	+12°	4000 ly	**86. NGC 6072**	−36°
0.3 ly	0.1′	10.7	2.0 ly	1.8′	12.1
2000 ly	**87. Abell 38**	−32°	4000 ly	**88. Abell 39**	+28°
1.5 ly	2.5′	20.0	3.5 ly	3′	15.6
6500 ly	**89. NGC 6210**	+24°	13,000 ly	**90. Abell 40**	−21°
1.4 ly	0.8′	11.7	2.2 ly	0.6′	18.0

Jul 27

PNe 91-100

#	PN Name	RA (H/M/S)	DEC (D/M/S)	Const.	Transit 9:00 PM	Transit 1:00 AM	Score	Class	Size (')	Distance (ly)	Diameter (ly)	Visual Mag
91	IC 4634	17h 01m 34s	-21° 49' 30"	Oph	Jul 27	May 27	2	BOY/y	0.20	13000	0.80	11.3
	Tiny bi-polar OIII-dominated PN with red ansae											
92	NGC 6302	17h 13m 44s	-37° 06' 12"	Sco	Jul 30	May 30	4	MRN	2.5	4000	3.0	10.1
	Butterfly — Multipolar outflow PN, with good results from SHO imaging											
93	NGC 6309	17h 14m 04s	-12° 54' 38"	Oph	Jul 30	May 31	2	BON/j	0.60	9000	1.5	11.5
	Box — Bipolar PN with visible jets											
94	NGC 6337	17h 22m 16s	-38° 29' 00"	Sco	Aug 1	Jun 2	4	BRY	1.0	5000	1.4	12.3
	Cheerio — Bipolar PN with bright torus ring and extended eye-shaped lobe region. Great inner region with progenitor and other stars. Similar to NGC6369											
95	Haro 2-08	17h 24m 46s	-21° 33' 36"	Oph	Aug 2	Jun 2	2	BCN	0.20	u	u	u
	Tiny PN resembling the much large Headphones Nebula											
96	Abell 41	17h 29m 04s	-15° 13' 21"	Ser	Aug 3	Jun 3	2	ERY/q	0.40	15000	1.7	16.0
	Smallest Abell PN in apparent size, and one of the furthest away from us. HII rim.											
97	NGC 6369	17h 29m 21s	-23° 45' 34"	Oph	Aug 3	Jun 3	3	BRY	1.2	3000	1.0	12.0
	Little Ghost — Bipolar PN, similar to NGC6337, with faint progenitor											
98	Abell 42	17h 31m 31s	-08° 19' 10"	Oph	Aug 4	Jun 4	1	ECY/r	1.0	11000	3.0	20.0
	The dimmest of all Abell PN which have a magnitude estimated. Faint progenitor star at the center.											
99	Kohoutek 1-14	17h 42m 36s	21° 30' 00"	Her	Aug 7	Jun 7	2	SOY/t	1.0	11000	3.0	u
	Showy OIII-dominant transparent PN with a wavy rim and visible progenitor											
100	Minkowski 1-28	17h 47m 36s	-22° 04' 00"	Sgr	Aug 8	Jun 8	3	BHN/c	0.90	15000	4.0	17.0
	Bipolar HII-dominated PN with blown out lobes											

u **95. Haro 2-8** -22°		15,000 ly **96. Abell 41** -15°
u 0.2' u		1.7 ly 0.4' 16.0
3000 ly **97. NGC 6369** -24°		11,000 ly **98. Abell 42** -8°
1.0 ly 1.2' 12.0		3.0 ly 1' 20.0
11,000 ly **99. Kohoutek 1-14** +22°		15,000 ly **100. Minkowski 1-28** -22°
3.0 ly 1.0' u		4 ly 0.9' 17.0

Aug 8 — PNe 101-110

#	PN Name	RA (H/M/S)	DEC (D/M/S)	Const.	Transit 9:00 PM	Transit 1:00 AM	Score	Class	Size (')	Distance (ly)	Diameter (ly)	Visual Mag
101	NGC 6445	17h 49m 15s	-20° 00' 36"	Sgr	Aug 8	Jun 8	3	MRN	2.5	4500	3.4	18.7
	Little Gem	Similar to NGC40, with white box that is cross-section of torus material from multi-polar PN gas flow. Viewing PN in side view										
102	Abell 43	17h 53m 32s	10° 37' 20"	Oph	Aug 9	Jun 10	3	SOY/ft	1.3	5000	2.0	15.0
		Progenitor visible. OIII-dominated with filamentary arcing structures weaving through the nebula.										
103	NGC 6543	17h 58m 33s	66° 37' 59"	Dra	Aug 11	Jun 11	5	ERN	5.0	3300	5.0	11.3
	Cat's Eye	Showcase Cat's Eye PN, with inner namesake eye and wispy hexagon-shaped outer extent										
104	NGC 6537	18h 05m 13s	-19° 50' 35"	Sgr	Aug 12	Jun 12	2	BHY/cf	2.0	5000	3.0	13.6
	Red Spider	Marvelous bi-polar 2 lobed shape of this HII PN, with rippled filaments and lobe breakthrough. Visible progenitor										
105	Minkowski 1-41	18h 09m 30s	-24° 12' 15"	Sgr	Aug 13	Jun 14	3	PHN/c	2.0	2800	1.8	15.0
		Multipolar HII-dominated PN										
106	NGC 6563	18h 12m 03s	-33° 52' 07"	Sgr	Aug 14	Jun 14	3	BRY/r	1.0	6000	1.9	11.0
	Southern Ring	M57-like bipolar PN, viewed end-on with bright torus ring, HII rim, OIII central portion, and jet created lobes										
107	NGC 6572	18h 12m 06s	06° 51' 13"	Oph	Aug 14	Jun 14	1	EON	0.25	5000	0.40	10.8
	Blue Racquetball	Tiny bipolar OIII PN with progenitor, 2 magnitudes brighter than M57 but 5 times smaller.										
108	Minkowski 4-09	18h 14m 18s	-04° 59' 22"	Ser	Aug 15	Jun 15	2	ERY	0.70	6000	1.3	16.6
		Elliptical ring PN with visible progenitor										
109	Sh2-068	18h 24m 58s	00° 51' 36"	Ser	Aug 17	Jun 17	3	ACY/i	8.0	1300	3.3	16.5
	Flaming Skull	Large faint PN with HII and OIII contribution and significant ISM interaction										
110	NGC 6629	18h 25m 42s	-23° 12' 10"	Sgr	Aug 17	Jun 18	2	BOY/r	0.40	7700	1.0	10.0
		Small bright OIII PN with asymmetric halo										

101. NGC 6445 — 4500 ly, -20°, 3.4 ly, 2.5', 18.7

102. Abell 43 — 5000 ly, +11°, 2 ly, 1.3', 15.0

103. NGC 6543 — 3300 ly, +67°, 5 ly, 5', 11.3

104. NGC 6537 — 5000 ly, -20°, 3 ly, 2', 13.6

PNe 111-120

Aug 19

#	PN Name	RA (H/M/S)	DEC (D/M/S)	Const.	Transit 9:00 PM	Transit 1:00 AM	Score	Class	Size (')	Distance (ly)	Diameter (ly)	Visual Mag
111	Abell 44	18h 30m 11s	-16° 45' 27"	Sgr	Aug 19	Jun 19	2	BRN/q	1.0	7000	2.0	16.0
	Bi-polar in nature and looking like a short cylinder. Strong HII rim, no progenitor star.											
112	Abell 45	18h 30m 17s	-11° 36' 54"	Sct	Aug 19	Jun 19	2	PHN/f	7.0	u	u	13.0
	Looks like Hii region. No progenitor or OIII signal. I could not confirm that this object is a PN.											
113	Abell 46	18h 31m 19s	26° 56' 17"	Lyr	Aug 19	Jun 19	2	EOY/o	1.3	7200	2.7	15.0
	Similar strengths of HII & OIII signal. Central progenitor star is bright, quite a contrast to the much fainter nebula.											
114	Minkowski 3-28	18h 32m 41s	-10° 05' 48"	Sct	Aug 19	Jun 19	3	BRN/c	0.40	16000	2.0	15.0
	Bipolar PN with both end lobes blown out											
115	Minkowski 3-55	18h 33m 15s	-10° 15' 07"	Sct	Aug 19	Jun 20	1	BRN/c	0.20	u	u	20.0
	Distant bipolar PN with blown out lobes											
116	Abell 47	18h 35m 22s	-00° 13' 32"	Ser	Aug 20	Jun 20	2	EHY/q	0.30	30000	2.6	19.0
	Most distant Abell PN and smallest in apparent size. Tiny, HII-dominated. Strong outer rim.											
117	Minkowski 1-57	18h 40m 20s	-10° 39' 47"	Sct	Aug 21	Jun 21	2	BRN/c	0.50	15000	2.3	14.0
	Tiny bipolar PN with blown out lobes											
118	Abell 48	18h 42m 49s	-03° 13' 00"	Aql	Aug 22	Jun 22	2	PHN	0.80	11800	2.7	17.0
	Nice diamond-shaped double ring structure. Strong HII signal. Progenitor star not likely seen.											
119	Minkowski 1-64	18h 50m 00s	35° 15' 00"	Lyr	Aug 24	Jun 24	2	EHN/r	0.40	18000	2.3	u
	White circular PN with HII rim											
120	Abell 49	18h 53m 29s	-06° 29' 14"	Sct	Aug 24	Jun 25	2	ERN/r	1.0	u	u	16.0
	HII with OIII central region. Some breakout seems to appear at 5 and 11 o'clock. No progenitor star.											

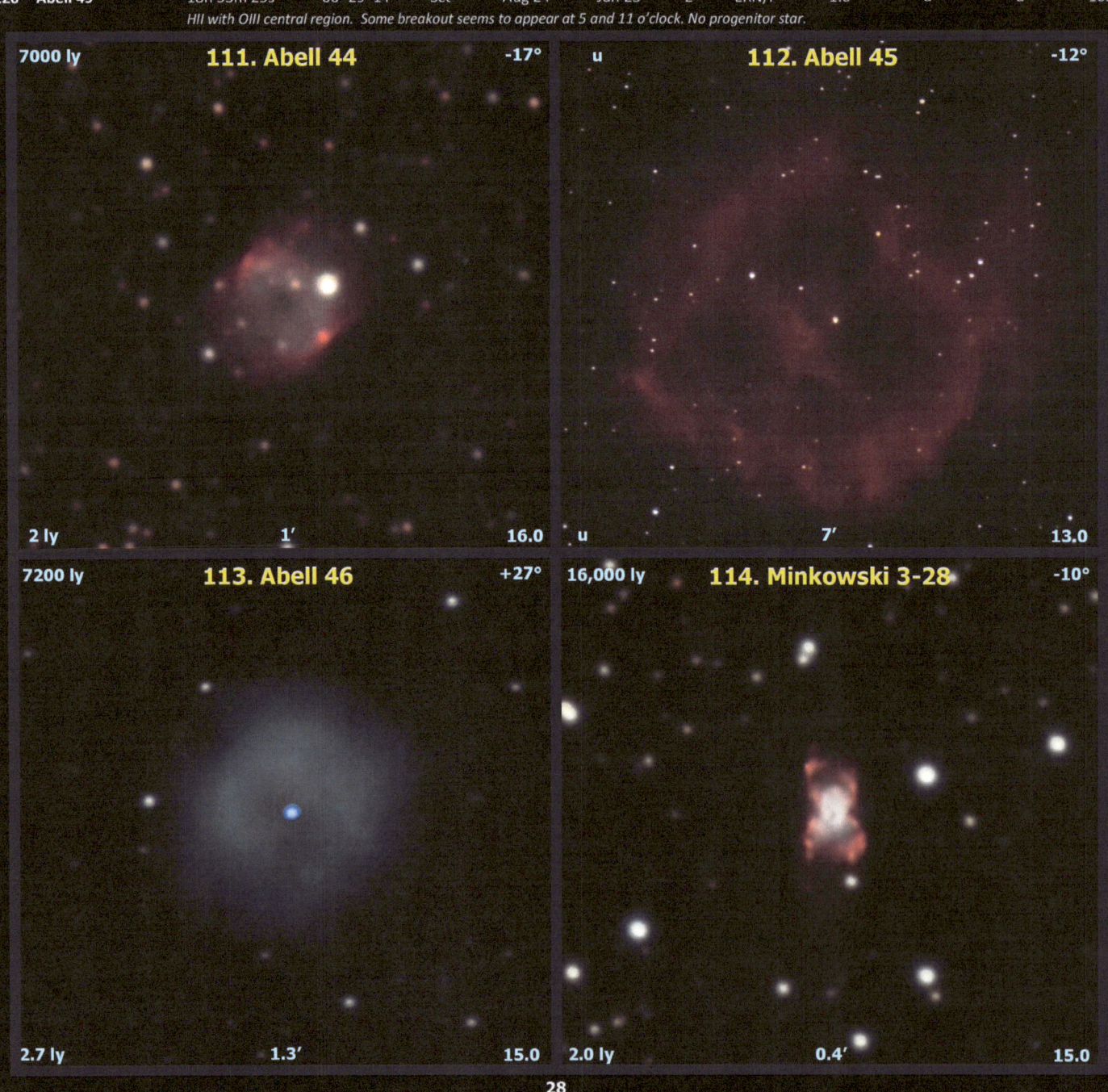

u	**115. Minkowski 3-55**	-10°	30,000 ly	**116. Abell 47**	-0°
u	0.2'	20.0	2.6 ly	0.3'	19.0
15,000 ly	**117. Minkowski 1-57**	-11°	11,800 ly	**118. Abell 48**	-3°
2.3 ly	0.5'	14.0	2.7 ly	0.8'	17.0
18,000 ly	**119. Minkowski 1-64**	+35°	u	**120. Abell 49**	-6°
2.3 ly	0.4'	u	u	1.0'	16.0

Aug 25 — PNe 121-130

#	PN Name	RA (H/M/S)	DEC (D/M/S)	Const.	Transit 9:00 PM	Transit 1:00 AM	Score	Class	Size (')	Distance (ly)	Diameter (ly)	Visual Mag
121	M 057 *Ring*	18h 53m 35s	33° 01' 44"	Lyr	Aug 25	Jun 25	5	BRY/r	1.5	2300	1.0	15.8
	There are many ring-shaped planetary nebulae in the sky but only one Ring Nebula.											
122	IC 1295	18h 54m 37s	-08° 49' 36"	Sct	Aug 25	Jun 25	3	SOY/ot	2.0	3400	2.0	12.5
	Delicate M97-type PN, strong OIII with voids and a thin rim, underappreciated PN											
123	YM 16	18h 54m 57s	06° 02' 31"	Ser	Aug 25	Jun 25	2	EHY/i	6.0	u	u	u
	Large elliptical HII PN with some texture											
124	Abell 50	18h 59m 20s	48° 27' 57"	Dra	Aug 26	Jun 26	2	SOY/o	0.50	17000	2.4	16.6
	One of smallest Abell PN in apparent size. Beautiful color and some structure is seen.											
125	Abell 51	19h 01m 01s	-18° 12' 16"	Sgr	Aug 26	Jun 27	2	SOY/r	1.0	6000	1.7	15.4
	OIII-dominated. The blue central progenitor star is clearly visible. The spherical nebula has a thick textured ring.											
126	Sh2-071	19h 02m 00s	02° 09' 11"	Aql	Aug 27	Jun 27	3	MHY	3.0	3000	2.5	14.0
	Multipolar HI-dominated PN with unusual, complex asymmetric shape											
127	NGC 6741 *Phantom Streak*	19h 02m 37s	-00° 26' 57"	Aql	Aug 27	Jun 27	1	ECN	0.15	12000	0.50	11.5
	Tiny PN with bright rim in Aquila											
128	HaTr 11	19h 02m 59s	03° 02' 21"	Aql	Aug 27	Jun 27	2	MRN	1.0	u	u	u
	Multipolar peculiar PN											
129	Kohoutek 1-17	19h 03m 36s	19° 12' 00"	Sge	Aug 27	Jun 27	3	BOY/r	0.80	10000	2.2	u
	OIII-dominating ring PN with like ISM interaction											
130	Abell 52	19h 04m 32s	17° 57' 10"	Aql	Aug 27	Jun 28	2	EOY/iq	1.0	6000	1.7	14.0
	Odd asymmetric structure. Bright rim section at top left. OIII-dominated. Progenitor star visible.											

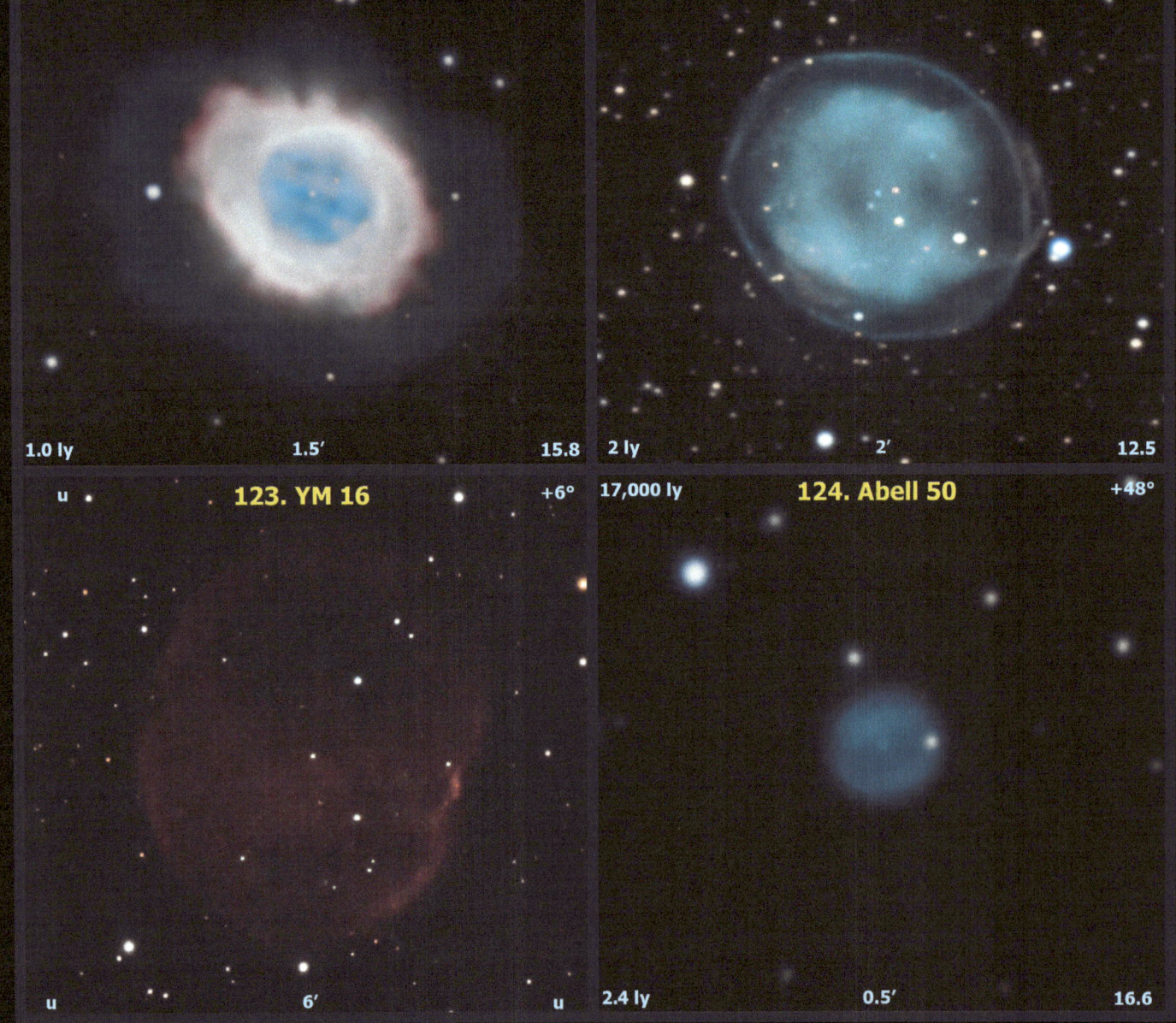

6000 ly	**125. Abell 51**	−18°	3000 ly	**126. Sh2-71**	+2°
1.7 ly	1.0′	15.4	2.4 ly	3′	16.0
12,000 ly	**127. NGC 6741**	−0°	u	**128. HaTr 11**	+3°
0.5 ly	0.15′	11.5	u	1′	u
10,000 ly	**129. Kohoutek 1-17**	+19°	6000 ly	**130. Abell 52**	+18°
2.2 ly	0.8′	u	1.7 ly	1′	14.0

PNe 131-140

Aug 28

#	PN Name	RA (H/M/S)	DEC (D/M/S)	Const.	Transit 9:00 PM	Transit 1:00 AM	Score	Class	Size (')	Distance (ly)	Diameter (ly)	Visual Mag
131	NGC 6751	19h 05m 56s	-05° 59' 31"	Aql	Aug 28	Jun 28	3	BRY/f	0.70	8400	1.8	15.5
	Dandelion Puff Ball	Complex PN with 5 different regions, this bipolar nebula has outflow resulting in multiple shells.										
132	Abell 53	19h 06m 46s	06° 23' 56"	Aql	Aug 28	Jun 28	2	SRN/r	0.50	6300	1.0	14.0
		Smallest Abell PN in actual diameter. HII-dominated rim of the nebula. Star superimposed on rim at 10 o'clock.										
133	Abell 54	19h 08m 39s	22° 58' 51"	Vul	Aug 28	Jun 29	1	ERY	1.0	9000	2.6	17.0
		Very dim. Hii rim while inner section is stronger in OIII. Star at center but not likely the progenitor star.										
134	Abell 55	19h 10m 30s	-02° 21' 02"	Aql	Aug 29	Jun 29	2	BRN/q	1.3	8200	3.0	13.0
		HII rim, strong along left & right edges. OIII strongest in central region. May be bipolar nebula with breakthrough.										
135	NGC 6765	19h 11m 07s	30° 32' 45"	Lyr	Aug 29	Jun 29	2	PON	1.0	5000	1.5	12.9
		Strangely shaped OIII PN.										
136	Abell 56	19h 13m 07s	02° 52' 49"	Aql	Aug 29	Jun 30	2	EHY/hr	3.0	4900	4.0	u
		Unusual in both color & structure. Wide, bright circular ring. Bi-polar, even multi-polar, structure seen.										
137	NGC 6772	19h 14m 36s	-02° 42' 24"	Aql	Aug 30	Jun 30	2	ERY	1.5	4000	1.7	16.8
		Squarish shaped and diffuse outer edges suggest ISM interaction. Progenitor visible.										
138	ETHOS 1	19h 16m 30s	36° 08' 56"	Lyr	Aug 30	Jul 1	3	BOY/fj	1.0	u	u	u
		Recently discovered bipolar PN. PN has both spherical and nested bipolar outflows. Progenitor visible.										
139	Abell 57	19h 17m 04s	25° 37' 26"	Vul	Aug 30	Jul 1	2	EOY	1.0	7000	2.0	17.7
		Odd, unique U-shape structure which is tough to interpret. OIII-dominated. Faint blue progenitor star is visible.										
140	Abell 58	19h 18m 20s	01° 46' 51"	Aql	Aug 31	Jul 1	2	EHY/hi	0.80	u	u	u
		Odd shape, perhaps due to ISM. "3" shape. Strong red color is due the dominant HII signal, with little OIII.										

131. NGC 6751 — 8400 ly, -6°, 1.8, 0.7', 15.5

132. Abell 53 — 6300 ly, +6°, 1.0 ly, 0.5', 14.0

133. Abell 54 — 9000 ly, +23°, 2.6 ly, 1', 17.0

134. Abell 55 — 8200 ly, -2°, 3.0 ly, 1.3', 13.0

Aug 31 — PNe 141-150

#	PN Name	RA (H/M/S)	DEC (D/M/S)	Const.	Transit 9:00 PM	Transit 1:00 AM	Score	Class	Size (')	Distance (ly)	Diameter (ly)	Visual Mag
141	NGC 6778 *Mini Dumbbell*	19h 18m 25s	-01° 35' 48"	Aql	Aug 31	Jul 1	2	BHY/c	0.50	10000	1.5	14.8
	Bipolar PN with lobes that have broken through. Resembles M76, which is 10x larger											
142	NGC 6781 *Snowglobe*	19h 18m 28s	06° 32' 20"	Aql	Aug 31	Jul 1	4	BHY	1.9	3000	1.8	11.8
	Bipolar PN just beginning to experience lobe breakout											
143	Abell 59	19h 18m 41s	19° 33' 56"	Sge	Aug 31	Jul 1	2	EHN/i	1.8	4600	2.4	16.0
	Looks like HII region, with bright red asymmetric arc at the top. Arc is not a typical PN structure.											
144	Abell 61	19h 19m 10s	46° 14' 36"	Cyg	Aug 31	Jul 1	2	SOY/t	3.0	3000	2.6	17.4
	Faint thin OIII-dominated rim around the outside edge of the top half of the nebula.											
145	Abell 60	19h 19m 17s	-12° 14' 52"	Sgr	Aug 31	Jul 1	2	EOY	1.7	5600	2.7	16.0
	OIII-dominated. May be bipolar, looking side-on into torus. Blown out bipolar regions extend left and right.											
146	Minkowski 4-14	19h 21m 01s	07° 36' 59"	Aql	Aug 31	Jul 2	1	BHN	0.20	25000	1.6	u
	Graceful HII-dominated bipolar PN with bright torus											
147	Kronberger 61 *Soccer Ball*	19h 21m 39s	38° 18' 57"	Lyr	Sep 1	Jul 2	2	SOY/t	5.0	13000	18	u
	Faint spherical PN											
148	DeHt 4	19h 26m 27s	13° 19' 35"	Aql	Sep 2	Jul 3	2	PHN/i	1.5	u	u	u
	Ear-shaped, HII-dominated											
149	NGC 6804 *Snowball*	19h 31m 35s	09° 13' 31"	Aql	Sep 3	Jul 4	2	BOY/r	2.5	2300	1.8	13.4
	OIII PN with bright torus and progenitor											
150	Abell 62	19h 33m 18s	10° 37' 01"	Aql	Sep 4	Jul 5	3	ERY/r	3.0	1600	1.4	15.0
	Strong HII hexagon rim while the inner section is stronger in OIII. PN is distorted. Small progenitor.											

141. NGC 6778 — 10,000 ly, -2°, 1.5, 0.5', 14.8

142. NGC 6781 — 3000 ly, +7°, 1.8 ly, 1.9', 11.8

143. Abell 59 — 4600 ly, +20°, 2.4 ly, 1.8', 16.0

144. Abell 61 — 3000 ly, +46°, 2.6 ly, 3', 17.4

Sep 4

PNe 151-160

#	PN Name	RA (H/M/S)	DEC (D/M/S)	Const.	Transit 9:00 PM	Transit 1:00 AM	Score	Class	Size (')	Distance (ly)	Diameter (ly)	Visual Mag
151	Campbell's Star	19h 34m 45s	30° 30' 59"	Cyg	Sep 4	Jul 5	2	XHY	0.20	10000	0.65	10.5
	Campbell's Star	Tiny stellar PN surrounding WC star (WC9-type) with strong hydrogen spectrum										
152	Kohoutek 2-07	19h 41m 25s	-20° 24' 44"	Sgr	Sep 6	Jul 7	1	AOY	2.5	6000	4.4	u
		Faint large ancient PN, faint progenitor visible										
153	PC 22	19h 42m 04s	13° 50' 35"	Aql	Sep 6	Jul 7	2	MON/c	1.0	17000	5.0	u
		Multipolar OIII-dominated PN with faint lobes										
154	Abell 63	19h 42m 10s	17° 05' 08"	Sge	Sep 6	Jul 7	2	PCY	0.70	9000	1.8	17.0
		Unique shape, like a broad X superimposed over circular shape. HII signal slightly stronger than OIII.										
155	NGC 6818	19h 43m 58s	-14° 09' 10"	Sgr	Sep 6	Jul 7	2	BOY/r	0.40	5000	0.60	9.3
	Little Gem	Bipolar OIII PN with white torus ring and central star										
156	Necklace Nebula	19h 43m 59s	17° 09' 02"	Sge	Sep 6	Jul 8	2	POY	0.50	15000	2.2	11.0
	Necklace	Tiny OIII-dominated PN with knots of gas around the ring										
157	NGC 6826	19h 44m 48s	50° 31' 30"	Cyg	Sep 7	Jul 8	4	BOY/r	0.50	4200	0.65	9.6
	Blinking Eye	OIII PN with visible progenitor and delicate faint outer ring										
158	Abell 65	19h 46m 34s	-23° 08' 12"	Sgr	Sep 7	Jul 8	3	EOY	4.0	4000	5.0	15.8
		Double-shelled bi-polar PN with binary central star. OIII-dominated inner shell oblong. Outer shell much fainter.										
159	Kohoutek 3-46	19h 50m 00s	33° 28' 00"	Cyg	Sep 8	Jul 9	2	BHN/c	1.0	u	u	18.9
		Nice bi-polar HII PN, viewed side-on. Both lobes broke through										
160	Minkowski 2-48	19h 50m 28s	25° 54' 22"	Vul	Sep 8	Jul 9	2	BHN/c	0.50	21000	3.0	u
		Unique bipolar HII-dominated PN with seemingly powerful jet influence										

151. Campbell's Star — 10,000 ly, +31°, 0.65, 0.2', 10.5

152. Kohoutek 2-7 — 6000 ly, -20°, 4.4 ly, 2.5', u

153. PC 22 — 17,000 ly, +14°, 5 ly, 1', u

154. Abell 63 — 9000 ly, +17°, 1.8 ly, 0.7', 17.0

5000 ly	**155. NGC 6818**	-14°	15,000 ly	**156. Necklace Nebula**	+17°
0.6 ly	0.4'	9.3	2.2 ly	0.5'	11.0
4200 ly	**157. NGC 6826**	+51°	4000 ly	**158. Abell 65**	-23°
0.7 ly	0.5'	9.6	5 ly	4'	15.8
u	**159. Kohoutek 3-46**	+33°	21,000 ly	**160. Minkowski 2-48**	+26°
u	1'	18.9	3 ly	0.5'	u

Sep 9 — PNe 161-170

#	PN Name	RA (H/M/S)	DEC (D/M/S)	Const.	Transit 9:00 PM	Transit 1:00 AM	Score	Class	Size (')	Distance (ly)	Diameter (ly)	Visual Mag
161	NGC 6842	19h 55m 02s	29° 17' 20"	Vul	Sep 9	Jul 10	2	EOY	1.0	4500	1.4	16.0
	OIII PN with progenitor and texture that looks like a ball of snakes											
162	Abell 66	19h 57m 32s	-21° 36' 37"	Sgr	Sep 10	Jul 11	2	ERY	4.5	2000	2.5	17.4
	HII rim. OIII irregular inner section. OIII strongest in 2 lobes, upper and lower. Some breakthrough at 12 o'clock.											
163	HDW 12	19h 58m 12s	-26° 28' 16"	Sgr	Sep 10	Jul 11	2	SHY	0.80	2100	0.50	18.5
	Faint asymmetric HII-dominated PN with offset progenitor											
164	Abell 67	19h 58m 29s	03° 02' 23"	Aql	Sep 10	Jul 11	2	EON/q	1.0	6500	2.0	18.0
	This PN is primarily comprised of OIII signal, resulting in its blue appearance. Progenitor not seen.											
165	Henize 1-4	19h 59m 18s	31° 33' 00"	Cyg	Sep 10	Jul 11	1	BHN	0.40	11000	1.4	u
	Small bright HII-dominated bipolar PN											
166	M 027 Dumbbell	19h 59m 36s	22° 43' 15"	Vul	Sep 10	Jul 11	5	BRY/c	15	1200	5.0	14.1
	Showcase PN with fascinating dumbbell-like structure with expanding halos											
167	Abell 68	20h 00m 11s	21° 42' 58"	Vul	Sep 10	Jul 12	2	ECY/q	0.70	6800	1.5	15.0
	HII and OIII signals are similar for this oval PN. Interesting texture in inner region, like an hourglass.											
168	NGC 6852	20h 00m 39s	01° 43' 41"	Aql	Sep 11	Jul 12	2	BOY	0.50	9000	1.4	17.5
	Similar to NGC6337. OIII PN with bright torus ring, extended eye-shaped lobe region, and visible progenitor.											
169	WeSb 5	20h 02m 42s	19° 36' 00"	Sge	Sep 11	Jul 12	2	SBN	2.5	u	u	17.6
	Very faint circular PN in Sagitta											
170	Minkowski 1-75	20h 04m 42s	31° 15' 00"	Cyg	Sep 12	Jul 13	3	MRN/c	0.30	14000	1.2	u
	Multipolar PN with bright white torus											

| 4500 ly | **161. NGC 6842** | +29° | 2000 ly | **162. Abell 66** | -22° |
| 1.4 | 1' | 16.0 | 2.5 ly | 4.5' | 17.4 |

| 2100 ly | **163. HDW 12** | -26° | 6500 ly | **164. Abell 67** | +3° |
| 0.5 ly | 0.8' | 18.5 | 2 ly | 1' | 18.0 |

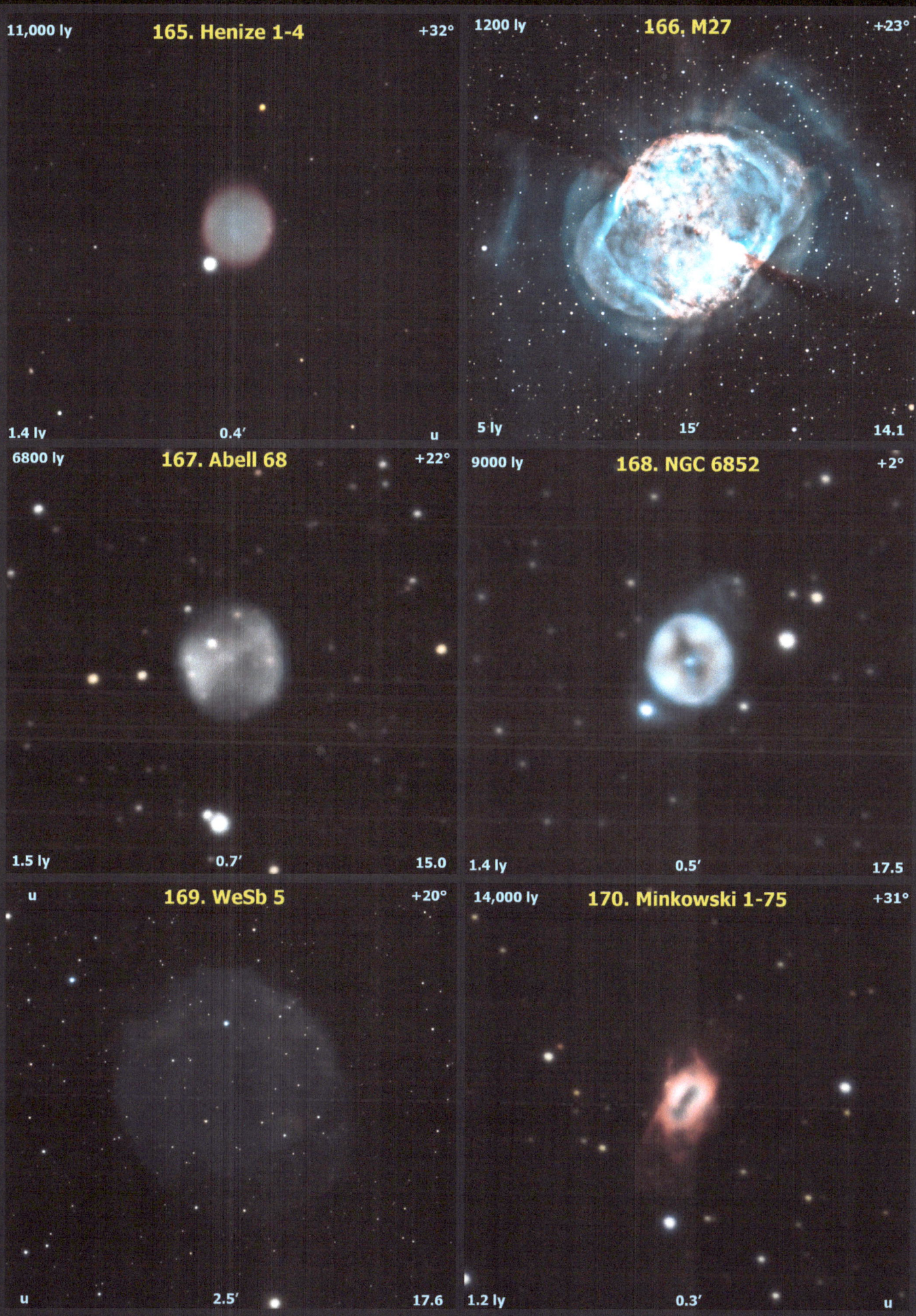

PNe 171-180

Sep 13

#	PN Name	RA (H/M/S)	DEC (D/M/S)	Const.	Transit 9:00 PM	Transit 1:00 AM	Score	Class	Size (')	Distance (ly)	Diameter (ly)	Visual Mag
171	Minkowski 4-17	20h 09m 00s	43° 44' 00"	Cyg	Sep 13	Jul 14	2	BRN/cr	1.0	u	u	u
	Bipolar PN with lobes and bright torus											
172	Weinberger 1-09	20h 09m 05s	26° 26' 56"	Vul	Sep 13	Jul 14	1	ERN	0.40	u	u	u
	Small HII PN ring of unknown distance											
173	NGC 6891	20h 15m 09s	12° 42' 16"	Del	Sep 14	Jul 15	2	EOY	0.30	12000	1.1	12.5
	Similar to NGC 7354, this OIII PN has dim white progenitor, bright elliptical inner football shell, and circular outer shell.											
174	Ju 1	20h 15m 26s	38° 02' 48"	Cyg	Sep 14	Jul 15	3	SON/t	4.0	4000	5.0	u
	Soap Bubble — Recently discovered faint PN with brighter rim											
175	NGC 6894	20h 16m 24s	30° 33' 55"	Cyg	Sep 15	Jul 16	2	BRY	0.70	5000	1.0	12.3
	Little Ring — Beautiful strong circular textured torus											
176	Abell 69	20h 19m 56s	38° 24' 31"	Cyg	Sep 15	Jul 17	2	EHN/r	0.40	14000	1.7	u
	HII rim. No progenitor. A faint outer halo is seen extending to about twice the diameter of the PN.											
177	NGC 6905	20h 23m 23s	20° 06' 16"	Del	Sep 16	Jul 17	2	BOY/fy	1.5	5750	2.5	14.5
	Blue Flash — Strong bipolar flow with ansae. Unusual football filamentary shape. Progenitor visible											
178	Abell 70	20h 31m 33s	-07° 05' 21"	Aql	Sep 18	Jul 20	2	ERY/hr	1.0	12000	3.5	15.0
	Diamond Ring — Visible central star. HII rim, OIII inner region. Whitish area of outer rim scalloped. Galaxy seen behind rim at 8 o'clock.											
179	Weinberger 1-10	20h 31m 54s	48° 53' 00"	Cyg	Sep 18	Jul 20	2	SBN/t	3.0	u	u	u
	Soap bubble like appearance for this PN with unknown distance											
180	Abell 71	20h 32m 23s	47° 21' 04"	Cyg	Sep 19	Jul 20	2	PHY/h	3.0	2400	2.0	19.3
	Could be PN or HII region. Nebula has spiral shape, with arms seen at the top and bottom.											

171. Minkowski 4-17 — +44° — u — 1' — u

172. Weinberger 1-9 — +26° — u — 0.4' — u

173. NGC 6891 — +13° — 12,000 ly — 1.1 ly — 0.3' — 12.5

174. Ju 1 (Soap Bubble) — +38° — 4000 ly — 5 ly — 4' — u

5000 ly	**175. NGC 6894**	+31°	14,000 ly	**176. Abell 69**	+38°
1 ly	0.7'	12.3	1.7 ly	0.4'	u

5700 ly	**177. NGC 6905**	+20°	12,000 ly	**178. Abell 70**	−7°
2.5 ly	1.5'	14.5	3.5 ly	1'	15.0

u	**179. Weinberger 1-10**	+49°	2400 ly	**180. Abell 71**	+47°
u	3'	u	2 ly	3'	19.3

Sep 21   PNe 181-190

#	PN Name	RA (H/M/S)	DEC (D/M/S)	Const.	Transit 9:00 PM	Transit 1:00 AM	Score	Class	Size (')	Distance (ly)	Diameter (ly)	Visual Mag
181	Kohoutek 4-53	20h 42m 18s	37° 24' 00"	Cyg	Sep 21	Jul 22	1	PHN	0.30	u	u	u
	Unusual PN shape, like a tiny Cat's Paw											
182	Kohoutek 4-55	20h 45m 12s	44° 24' 00"	Cyg	Sep 22	Jul 23	3	BRN	0.50	4500	0.70	16.5
	Beautiful small bipolar PN with point symmetric wings											
183	Abell 72	20h 50m 02s	13° 33' 28"	Del	Sep 23	Jul 24	3	BOY/f	3.0	3700	3.0	16.0
	OIII-dominated filamentary structure. Oblong nebula, likely a bi-polar PN. Faint galaxy at 6 o'clock.											
184	Ear Nebula	20h 50m 13s	46° 55' 00"	Cyg	Sep 23	Jul 24	3	PRN/fi	6.0	u	u	u
	Ear											
	Recently discovered complex faint PN											
185	Abell 73	20h 56m 26s	57° 25' 56"	Cep	Sep 25	Jul 26	2	ERY/q	1.4	6000	2.4	17.0
	HII rim and OIII inner section. PN distorted. 2 point-symmetric lobes of HII. Very faint progenitor.											
186	NGC 7008	21h 00m 33s	54° 32' 35"	Cyg	Sep 26	Jul 27	4	PBY	1.8	3000	1.5	12.0
	Fetus											
	Despite its nickname, this PN is in its final stage of life. Many of the "stars" seen here are actually bright condensations											
187	NGC 7009	21h 04m 11s	-11° 21' 50"	Aql	Sep 27	Jul 28	4	BOY/y	0.70	4500	1.0	12.0
	Saturn											
	Small Saturn-like OIII PN with red ansae											
188	NGC 7026	21h 06m 19s	47° 51' 08"	Cyg	Sep 27	Jul 28	4	MRY/f	0.70	10000	2.0	15.0
	Cheeseburger											
	Excellent multipolar PN with intact lobes and faint progenitor											
189	NGC 7027	21h 07m 02s	42° 14' 10"	Cyg	Sep 27	Jul 29	2	PBN	0.25	3000	0.22	8.8
	Magic Carpet											
	Small young PN with peculiar shape											
190	Weinberger 1-11	21h 10m 54s	50° 30' 00"	Cyg	Sep 28	Jul 30	1	EHN/r	0.40	u	u	u
	Elliptical HII PN of unknown distance											

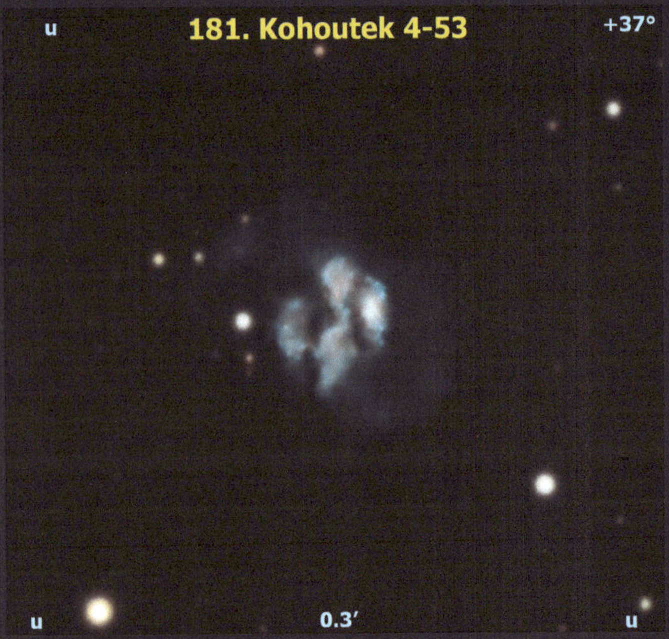

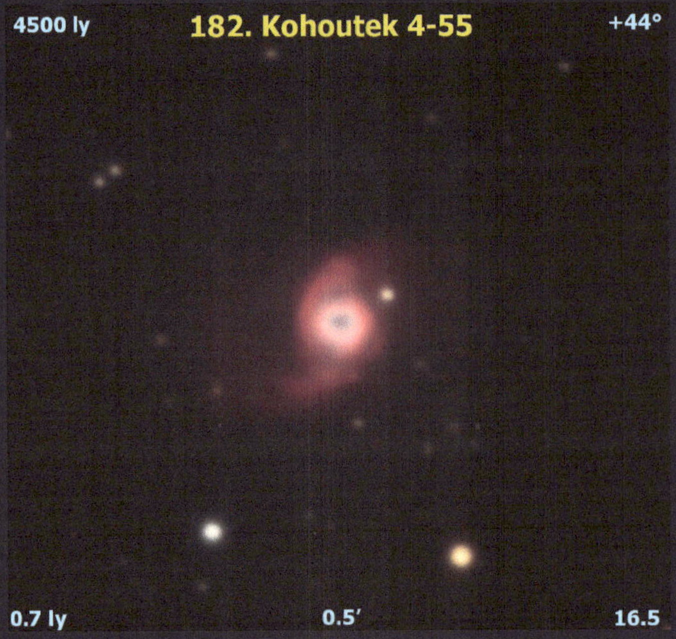

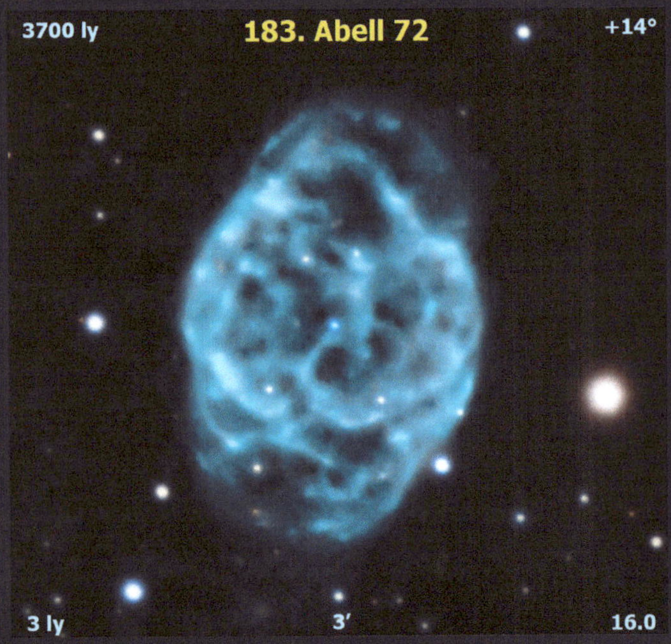

181. Kohoutek 4-53 — u / +37° / u / 0.3' / u

182. Kohoutek 4-55 — 4500 ly / +44° / 0.7 ly / 0.5' / 16.5

183. Abell 72 — 3700 ly / +14° / 3 ly / 3' / 16.0

184. Ear Nebula — u / +47° / u / 6' / u

6000 ly	**185. Abell 73**	+57°
2.4 ly	1.4'	17.0

3000 ly	**186. NGC 7008**	+55°
1.5 ly	1.8'	12.0

4500 ly	**187. NGC 7009**	+31°
1 ly	0.7'	12.0

10,000 ly	**188. NGC 7026**	+48°
2.0 ly	0.7'	15.0

3000 ly	**189. NGC 7027**	+42°
0.2 ly	0.3'	8.8

u	**190. Weinberger 1-11**	+51°
u	0.4'	u

PNe 191-200

Sep 29

#	PN Name	RA (H/M/S)	DEC (D/M/S)	Const.	Transit 9:00 PM	Transit 1:00 AM	Score	Class	Size (')	Distance (ly)	Diameter (ly)	Visual Mag
191	Sh1-089	21h 14m 06s	47° 46' 00"	Cyg	Sep 29	Jul 30	3	BRN/c	3.0	6400	6.0	u
	Bipolar HII-dominated PN with some lobe breakout											
192	NGC 7048	21h 14m 15s	46° 17' 18"	Cyg	Sep 29	Jul 30	4	BRY/c	1.0	5000	1.5	12.1
	Disk Ghost	*Bipolar PN, HII rim and OIII interior, with lobe breakout starting and faint progenitor visible*										
193	Abell 74	21h 16m 52s	24° 08' 51"	Vul	Sep 30	Jul 31	2	ARY/r	9.0	2200	5.7	17.1
	Faint progenitor star. HII rim with OII inner section. Large PN size.											
194	Weinberger 2-245	21h 18m 06s	43° 30' 00"	Cyg	Sep 30	Jul 31	1	PHN	1.2	u	u	u
	Very faint HII complex PN of unknown distance											
195	MWP 1	21h 17m 08s	34° 12' 27"	Cyg	Sep 30	Aug 1	2	AOY/c	13	4500	17	12.3
	Huge ancient PN with OIII-dominated signal and faint lobe arcs											
196	Abell 75	21h 26m 24s	62° 53' 28"	Cep	Oct 2	Aug 2	2	EOY	1.0	6000	1.7	15.0
	Progenitor star visible. Nebula squashed due to impact from ISM, which has affected the inner bright torus.											
197	Humason 1-2	21h 33m 06s	39° 38' 00"	Cyg	Oct 4	Aug 4	3	PON/y	0.50	9000	1.4	12.0
	Baby Dumbbell	*Beautiful small complex peculiar blue PN*										
198	Abell 78	21h 35m 29s	31° 41' 45"	Cyg	Oct 5	Aug 5	2	PRY/f	2.0	2300	1.3	13.0
	Odd structure - appears to be two major gas ejections, resulting in bright knots and filaments. Central star.											
199	NGC 7094	21h 36m 53s	12° 47' 19"	Peg	Oct 5	Aug 5	3	EOY/f	1.8	5400	2.8	13.7
	OIII PN with filaments and bright blue progenitor											
200	Minkowski 1-79	21h 37m 02s	48° 56' 06"	Cyg	Oct 5	Aug 5	3	MRN/c	0.80	8600	2.0	19.0
	Multipolar PN with burst out lobes and strong ISM influence											

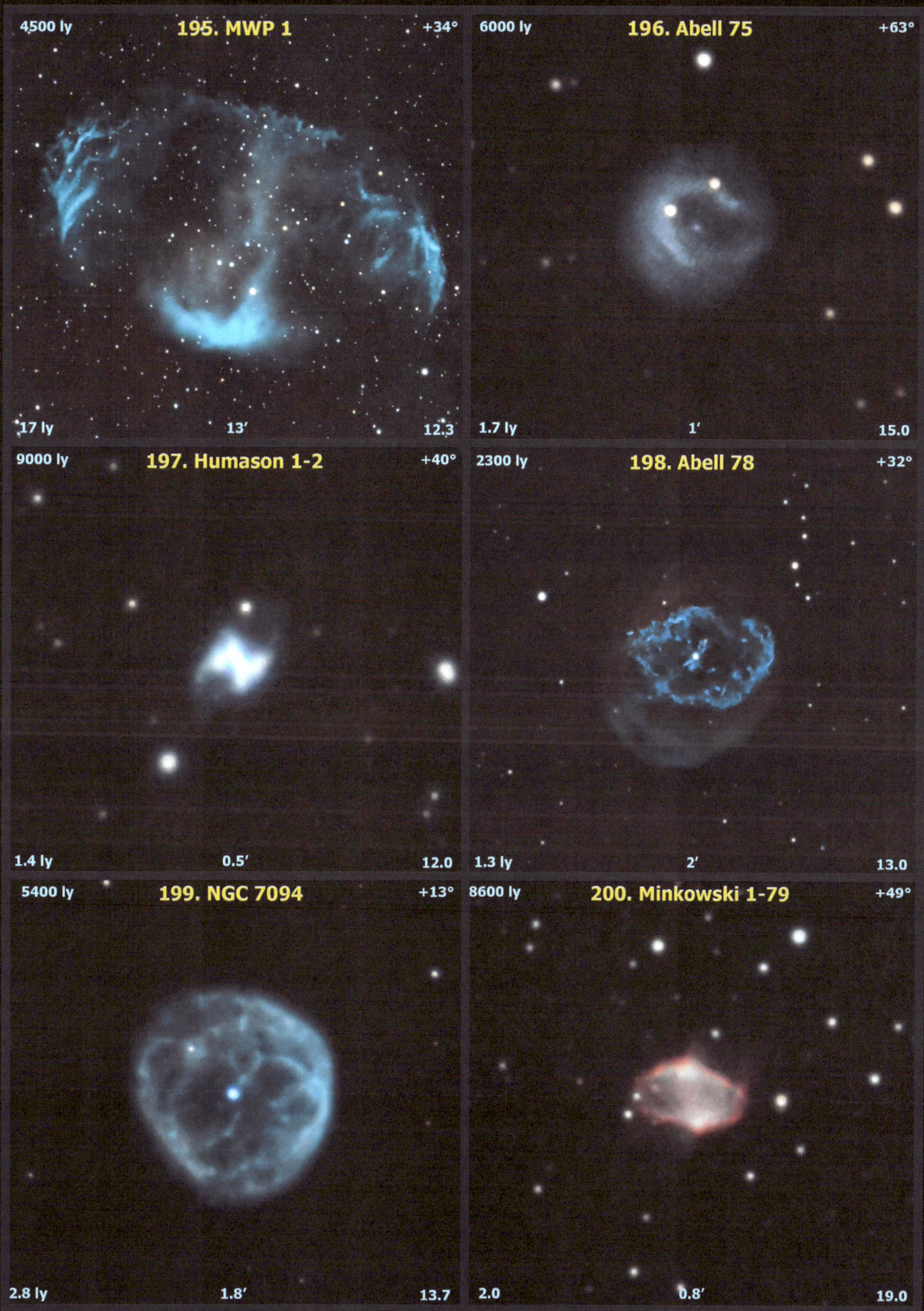

Oct 6 — PNe 201-210

#	PN Name	RA (H/M/S)	DEC (D/M/S)	Const.	Transit 9:00 PM	Transit 1:00 AM	Score	Class	Size (')	Distance (ly)	Diameter (ly)	Visual Mag
201	G100.4+04.6	21h 40m 00s	58° 59' 00"	Cep	Oct 6	Aug 6	3	PRY/y	1.0	u	u	u
	Multi-polar PN with HII outflows.											
202	NGC 7139	21h 46m 09s	63° 47' 30"	Cep	Oct 7	Aug 7	2	ERN/h	1.0	4000	1.3	13.3
	Hexagon shaped elliptical PN											
203	IC 5148	21h 59m 35s	-39° 23' 09"	Gru	Oct 11	Aug 11	4	ECY/r	2.0	3000	1.8	16.5
	Spare Tire — *Beautiful textured white circular ring with progenitor*											
204	IsWe 2	22h 13m 00s	65° 54' 00"	Cep	Oct 14	Aug 14	1	AHY	20	850	5.0	18.2
	Faint large ancient HII-dominated PN, one of the oldest PN at 50,000 years											
205	Minkowski 2-51	22h 16m 03s	57° 28' 31"	Cep	Oct 15	Aug 15	2	BRN/r	1.0	6000	1.8	u
	Bipolar PN in faint ring shape											
206	DeHt 5	22h 19m 34s	70° 56' 01"	Cep	Oct 16	Aug 16	2	ARY/i	9.0	1000	2.8	15.5
	Very faint, very old planetary nebula. Despite its PN characteristics, there is debate whether this is actually a PN.											
207	Minkowski 2-52	22h 20m 31s	57° 36' 18"	Cep	Oct 16	Aug 16	3	MRN	0.50	15000	2.2	14.0
	Unique Texas-shaped PN, likely multipolar											
208	IC 5217	22h 23m 56s	50° 58' 00"	Lac	Oct 17	Aug 17	1	XON	0.10	18000	0.6	11.3
	Tiny stellar OIII-dominated PN											
209	Abell 79	22h 26m 17s	54° 49' 41"	Lac	Oct 17	Aug 18	3	PHY/f	2.0	4000	2.3	17.0
	Unique looping PN structure. HII dominated. Arcing gas fronts and tiny lobes. Progenitor visible.											
210	NGC 7293	22h 29m 38s	-20° 50' 13"	Aql	Oct 18	Aug 18	5	BRY/r	12	650	2.5	13.5
	Helix — *Showcase Helix nebula, one of the closest PN to earth. Radial filaments point back to progenitor*											

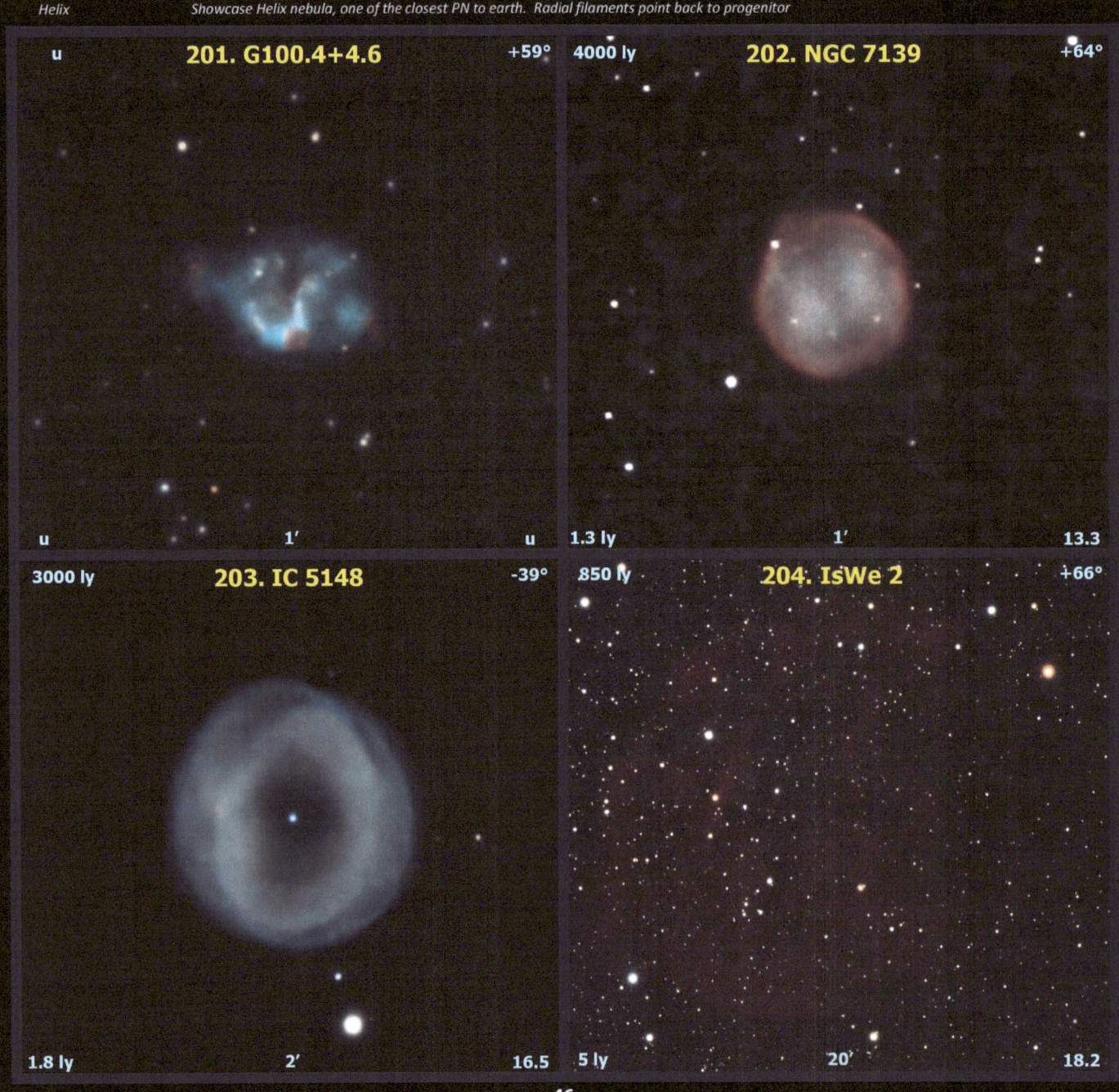

6000 ly	**205. Minkowski 2-51**	+57°	1000 ly	**206. DeHt 5**	+71°
1.8 ly	1′	u	2.8 ly	9′	15.5

15,000 ly	**207. Minkowski 2-52**	+58°	18,000 ly	**208. IC 5217**	+51°
2.2 ly	0.5′	14.0	0.6 ly	0.1′	11.3

4000 ly	**209. Abell 79**	+55°	650 ly	**210. NGC 7293**	−21°
2.3 ly	2′	17.0	2.5	12′	13.5

Oct 20   **PNe 211-220**

#	PN Name	RA (H/M/S)	DEC (D/M/S)	Const.	Transit 9:00 PM	Transit 1:00 AM	Score	Class	Size (')	Distance (ly)	Diameter (ly)	Visual Mag
211	Abell 80	22h 34m 46s	52° 26' 04"	Lac	Oct 20	Aug 20	2	EHY/q	2.2	6000	3.8	16.0
	HII rim with faint OIII inner region. Shape is not a ring but more like a short tube. Faint progenitor.											
212	NGC 7354	22h 40m 20s	61° 17' 07"	Cep	Oct 21	Aug 21	3	EOY	0.40	4200	0.50	12.2
	Similar to NGC 6891, this OIII PN has dim white progenitor, bright elliptical inner football shell, and circular outer shell.											
213	Abell 81	22h 42m 25s	80° 26' 33"	Cep	Oct 22	Aug 22	2	SOY/r	0.70	14000	2.8	14.0
	Beautiful OIII donut shape. Most northern of all Abell PNs. The central star is barely visible.											
214	KjPn 8	23h 24m 00s	60° 57' 00"	Cas	Nov 1	Sep 1	4	MHY/fy	11	6000	20	u
	Unique multipolar HII-dominated PN with fantastic structure											
215	NGC 7662 *Blue Snowball*	23h 25m 54s	42° 32' 06"	And	Nov 2	Sep 2	3	EHY/y	0.50	6400	1.0	8.3
	Similar to NGC6826, this is a bi-polar OIII nebula with torus and filaments, surrounded by faint outer sphere											
216	Minkowski 2-55	23h 31m 54s	70° 22' 00"	Cep	Nov 3	Sep 3	3	BRN/o	1.0	7800	2.2	u
	Bipolar HII-dominated PN with interesting inner void structure											
217	Jones 1	23h 35m 53s	30° 28' 06"	Peg	Nov 4	Sep 4	3	EOY/iq	5.0	2600	4.0	15.6
	Lovely OIII-dominated elliptical PN with progenitor and some texture											
218	Kohoutek 1-20	23h 39m 10s	48° 12' 31"	And	Nov 5	Sep 5	2	EON/r	0.70	14000	2.7	16.5
	OIII-dominated elliptical PN with ring appearance and no progenitor											
219	Abell 82	23h 45m 47s	57° 04' 01"	Cas	Nov 7	Sep 7	2	ERN/q	1.7	6500	3.2	13.0
	Looks like bi-polar PN with axis running from upper left to lower right. Torus section is constricting the jet flow. No progenitor.											
220	Abell 83	23h 46m 46s	54° 44' 40"	Cas	Nov 7	Sep 7	1	ERN/h	0.70	u	u	15.0
	Faint. HII rim with OIII inner section. No progenitor star.											

211. Abell 80 — 6000 ly, +52°, 3.8 ly, 2.2', 16.0

212. NGC 7354 — 4200 ly, +61°, 0.5 ly, 0.4', 12.2

213. Abell 81 — 14,000 ly, +80°, 2.8 ly, 0.7', 14.0

214. KjPn 8 (Probable PN) — 6000 ly, +61°, 20 ly, 11', u

| 6400 ly | **215. NGC 7662** | +43° | 7800 ly | **216. Minkowski 2-55** | +70° |
| 1 ly | 0.5′ | 8.3 | 2.2 ly | 1′ | u |

| 2600 ly | **217. Jones 1** | +30° | 14,000 ly | **218. Kohoutek 1-20** | +48° |
| 4 ly | 5′ | 15.6 | 2.7 ly | 0.7′ | 16.5 |

| 6500 ly | **219. Abell 82** | +57° | u | **220. Abell 83** | +55° |
| 3.2 ly | 1.7′ | 13.0 | u | 0.7′ | 15.0 |

PNe 221-230

Nov 7

#	PN Name	RA (H/M/S)	DEC (D/M/S)	Const.	Transit 9:00 PM	Transit 1:00 AM	Score	Class	Size (')	Distance (ly)	Diameter (ly)	Visual Mag
221	Sh2-174 *Valentine Rose*	23h 47m 08s	80° 49' 22"	Cep	Nov 7	Sep 7	4	ACY/i	17	1000	5.0	14.7
	Faint ancient PN, unusually asymmetric, with both HII and OIII signal											
222	Abell 84	23h 47m 45s	51° 23' 58"	Cas	Nov 7	Sep 7	2	ERY/hq	2.7	5000	4.0	18.6
	HII rim with OIII inner region. A bluish central star is faintly visible. Some subtle structure is evident in the nebula.											
223	Abell 86	00h 01m 33s	70° 42' 42"	Cep	Nov 11	Sep 11	2	SHN/i	1.3	7300	3.0	u
	Dominant HII. No progenitor star.											
224	Sh1-118	00h 07m 20s	64° 57' 21"	Cas	Nov 12	Sep 12	1	EHN	3.0	4000	3.4	13.9
	Identified as a PN but structure looks more like a HII region											
225	Abell 01	00h 12m 36s	69° 10' 41"	Cep	Nov 13	Sep 14	2	SHN/h	0.80	8200	2.0	19.0
	HII much stronger than OIII. Outer edge of PN resembles hexagon. No progenitor star.											
226	NGC 0040 *Bowtie*	00h 13m 01s	72° 31' 19"	Cep	Nov 13	Sep 14	4	BHY/cf	0.60	3500	0.60	11.5
	Bright small HII bipolar PM with filaments and bright progenitor											
227	Böhm-Vitense 5-1	00h 19m 59s	62° 59' 06"	Cas	Nov 15	Sep 15	2	PCN/fi	1.5	7200	3.0	u
	Oddly shaped PN with bright linear arcs											
228	OU 2	00h 30m 57s	61° 24' 34"	Cas	Nov 18	Sep 18	2	EOY/r	1.2	u	u	u
	Recently discovered OIII-dominated faint ring PN											
229	Weinberger 1-01	00h 38m 54s	66° 23' 49"	Cas	Nov 20	Sep 20	1	EHN	0.30	19000	1.6	u
	Tiny hexagon shaped HII PN											
230	Böhm-Vitense 5-2 *Sh2-179*	00h 40m 24s	62° 51' 00"	Cas	Nov 20	Sep 21	3	PRN/i	0.70	650	0.14	21.5
	Small beautiful PN with unique shape, cut off by a dark nebula											

221. Sh2-174 — 1000 ly, +81°, 5 ly, 17', 14.7

222. Abell 84 — 5000 ly, +51°, 4 ly, 2.7', 16.6

223. Abell 86 — 7300 ly, +71°, 3 ly, 1.3', u

224. Sh1-118 (Probable PN) — 4000 ly, +65°, 3.4 ly, 3', 13.9

8200 ly	**225. Abell 1**	+69°
2 ly	0.8′	19.0

3500 ly	**226. NGC 40**	+73°
0.6 ly	0.6′	11.5

7200 ly	**227. Böhm-Vitense 5-1**	+31°
3 ly	1.5′	u

u	**228. OU 2**	+61°
u	1.2′	u

19,000 ly	**229. Weinberger 1-1**	+66°
1.6 ly	0.3′	u

650 ly	**230. Böhm-Vitense 5-2**	+63°
0.14 ly	0.7′	21.5

Nov 22   **PNe 231-240**

#	PN Name	RA (H/M/S)	DEC (D/M/S)	Const.	Transit 9:00 PM	Transit 1:00 AM	Score	Class	Size (')	Distance (ly)	Diameter (ly)	Visual Mag
231	Abell 02	00h 45m 36s	57° 57' 24"	Cas	Nov 22	Sep 22	2	SOY/ho	0.50	13000	2.0	15.0
	Central star barely visible. Structure similar to M97 – round, OIII, 2 dark circular inner region voids.											
232	NGC 0246	00h 47m 03s	-11° 52' 19"	Cet	Nov 22	Sep 22	4	BOY/o	4.5	1600	2.2	11.8
	Skull											
	OIII PN with central owl-type void regions and slight bipolar breakthrough, progenitor visible											
233	EGB 01	01h 07m 06s	73° 33' 00"	Cas	Nov 27	Sep 27	2	ARY/i	3.5	1000	1.0	16.4
	Deformed elliptical PN with progenitor visible and some likely ISM influence											
234	Sh2-188	01h 30m 33s	58° 24' 51"	Cas	Dec 3	Oct 3	4	ARY/fi	9.0	850	2.1	17.4
	Dolphin											
	Beautiful filamentary PN and asymmetric shape											
235	M 076	01h 42m 18s	51° 34' 15"	Per	Dec 6	Oct 6	4	BRN	3.0	3000	2.7	17.5
	Little Dumbbell											
	Beautiful PN, Dimmest object in the Messier catalogue											
236	Böhm-Vitense 5-3	01h 53m 03s	56° 24' 00"	Per	Dec 9	Oct 9	2	ERN/r	0.50	23000	3.5	15.0
	Small bright PN, no progenitor.											
237	Ferrero 6	01h 56m 26s	65° 28' 20"	Cas	Dec 10	Oct 10	2	ERY	3.3	u	u	u
	Faint PN, no distance known. Progenitor visible.											
238	IC 1747	01h 57m 36s	63° 19' 18"	Cas	Dec 10	Oct 10	2	EOY	0.20	10000	0.60	15.4
	Holepunch											
	Small with nice inner detail, OIII dominated, progenitor seen.											
239	Kohoutek 3-91	01h 58m 36s	66° 34' 00"	Cas	Dec 10	Oct 10	1	XHN	0.30	28000	2.5	21.0
	One of the furthest PN at a distance of almost 30,000 light years											
240	Kohoutek 3-92	02h 03m 40s	64° 57' 36"	Cas	Dec 12	Oct 12	2	BRN/c	0.30	22000	2.0	17.0
	Bipolar PN with some lobe breakout											

231. Abell 2 — 13,000 ly — +58° — 2 ly — 0.5' — 15.0

232. NGC 246 — 1600 ly — -12° — 2.2 ly — 4.5' — 11.8

233. EGB 1 — 1000 ly — +74° — 1 ly — 3.5' — 16.4

234. Sh2-188 — 850 ly — +58° — 2.1 ly — 9' — 17.4

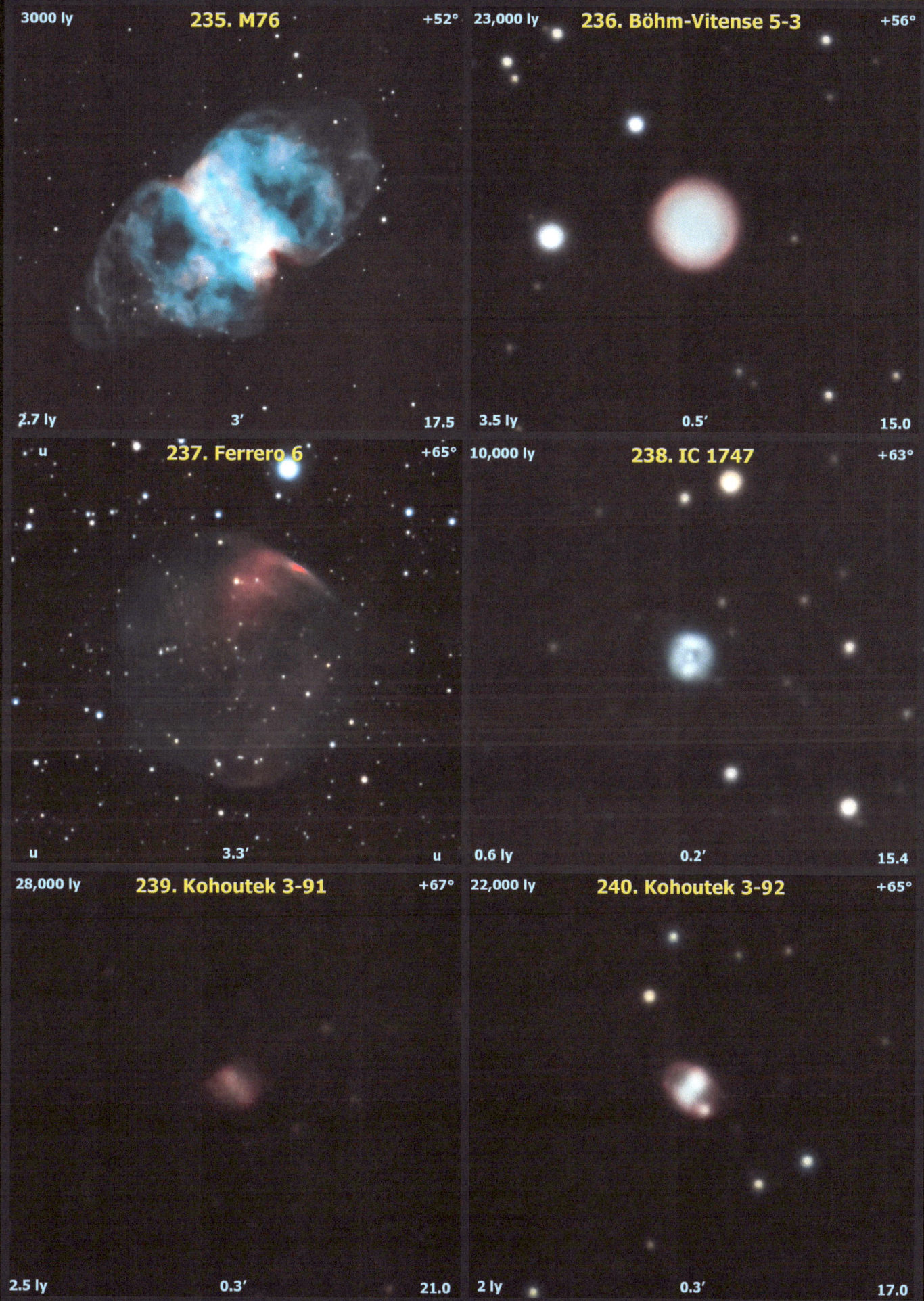

Dec 14

PNe 241-250

#	PN Name	RA (H / M / S)	DEC (D / M / S)	Const.	Transit 9:00 PM	Transit 1:00 AM	Score	Class	Size (')	Distance (ly)	Diameter (ly)	Visual Mag
241	Abell 03	02h 12m 12s	64° 09' 05"	Cas	Dec 14	Oct 14	2	SHY/hi	1.2	8400	3.0	16.0
	Progenitor star clearly visible. Nebula is not quite round - perhaps interaction between nebula and ISM.											
242	G132.8+2.0	02h 20m 45s	63° 11' 34"	Cas	Dec 16	Oct 16	2	ERN	0.50	4300	0.70	u
	Faint PN with interesting texture, next to bright star											
243	Kohoutek 3-93	02h 26m 30s	65° 47' 53"	Cas	Dec 17	Oct 18	1	XRN	0.20	u	u	18.0
	Tiny dim PN with rim stronger in HII. No progenitor.											
244	WeBo 1	02h 40m 14s	61° 09' 17"	Cas	Dec 21	Oct 21	2	EHY	1.3	5120	2.0	14.5
	Interesting elliptical HII ring PN											
245	Abell 04	02h 45m 26s	42° 32' 36"	Per	Dec 22	Oct 22	2	SOY/o	0.40	18000	2.0	15.0
	Central star barely visible. Structure similar to M97 – round, OIII, dark inner voids. Many galaxies in background.											
246	Abell 05	02h 52m 13s	50° 35' 52"	Per	Dec 24	Oct 24	2	EHN/q	2.5	5400	4.0	u
	Mainly HII. No central star. Oval banded shape.											
247	Abell 06	02h 58m 53s	64° 29' 59"	Cas	Dec 26	Oct 26	2	SCY/t	3.0	3100	2.7	15.0
	Mainly OIII. Faint central star. Bright rim.											
248	HFG 1	03h 01m 00s	64° 58' 00"	Cas	Dec 26	Oct 26	3	AOY	8.0	2300	5.0	14.6
	Large ancient OIII PN in field of HII, adjacent to Abell 6											
249	IC 0289	03h 10m 19s	61° 19' 01"	Cas	Dec 28	Oct 29	2	BCY	0.75	5000	1.1	16.0
	Small, interesting PN with disrupted bright white torus and visible progenitor.											
250	Sh2-200	03h 11m 01s	62° 47' 45"	Cas	Dec 29	Oct 29	4	AON/io	6.0	3600	6.0	u
	Large faint ancient PN, with nice texture and OIII-dominated throughout											

8400 ly	**241. Abell 3**	+64°	4300 ly	**242. G132.8+2.0**	+63°
3 ly	1.2'	16.0	0.7 ly	0.5'	u

u	**243. Kohoutek 3-93**	+66°	5120 ly	**244. WeBo 1**	+61°
u	0.2'	18.0	2 ly	1.3'	14.5

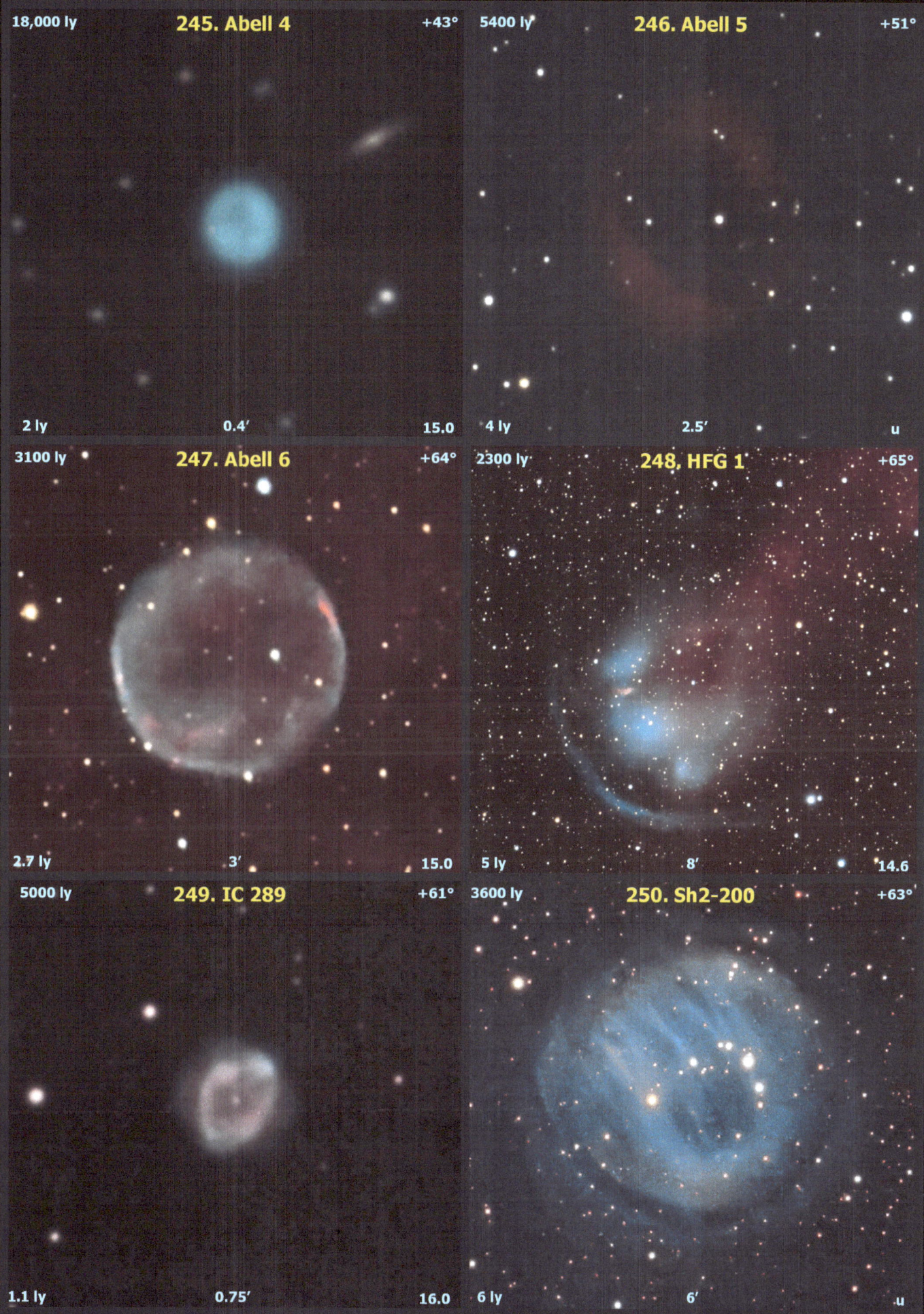

PNe Morphology

PNe are complex and challenging to understand – thousands of papers have been written on these beautiful objects over the past 50 years. We don't yet fully understand the underlying mechanics with certainty. Each year, scientists learn more about them and pass this knowledge on to us.

A PN is often simply described as a shell of ionized gas ejected from a red giant star late in life. But the actual mechanics are much more complex than that. PNe come in a seemingly infinite array of shapes, sizes and colors.

It is now believed that the primary reason for this variety is the nature of the progenitor (source) star. Although many different categorizations have been proposed over the years, here I am presenting a simple 4 category view – spherical, bipolar, elliptical, and peculiar, as described on this and the following 3 pages.

PNe Morphology - Spherical

About 10% of PNe are believed to be spherical. Because of the symmetry, the progenitor in these cases is believed to be a single star (no companion). Although this subject is still being debated, it is believed that single stars do not have the required rotational speeds late in their life to form the bipolar and elliptical PNe described further below.

Spherical PNe have little internal structure, a faint (if any) rim, and a circular projection. The inner region can either be clear or somewhat occluded, as seen in the examples shown below:

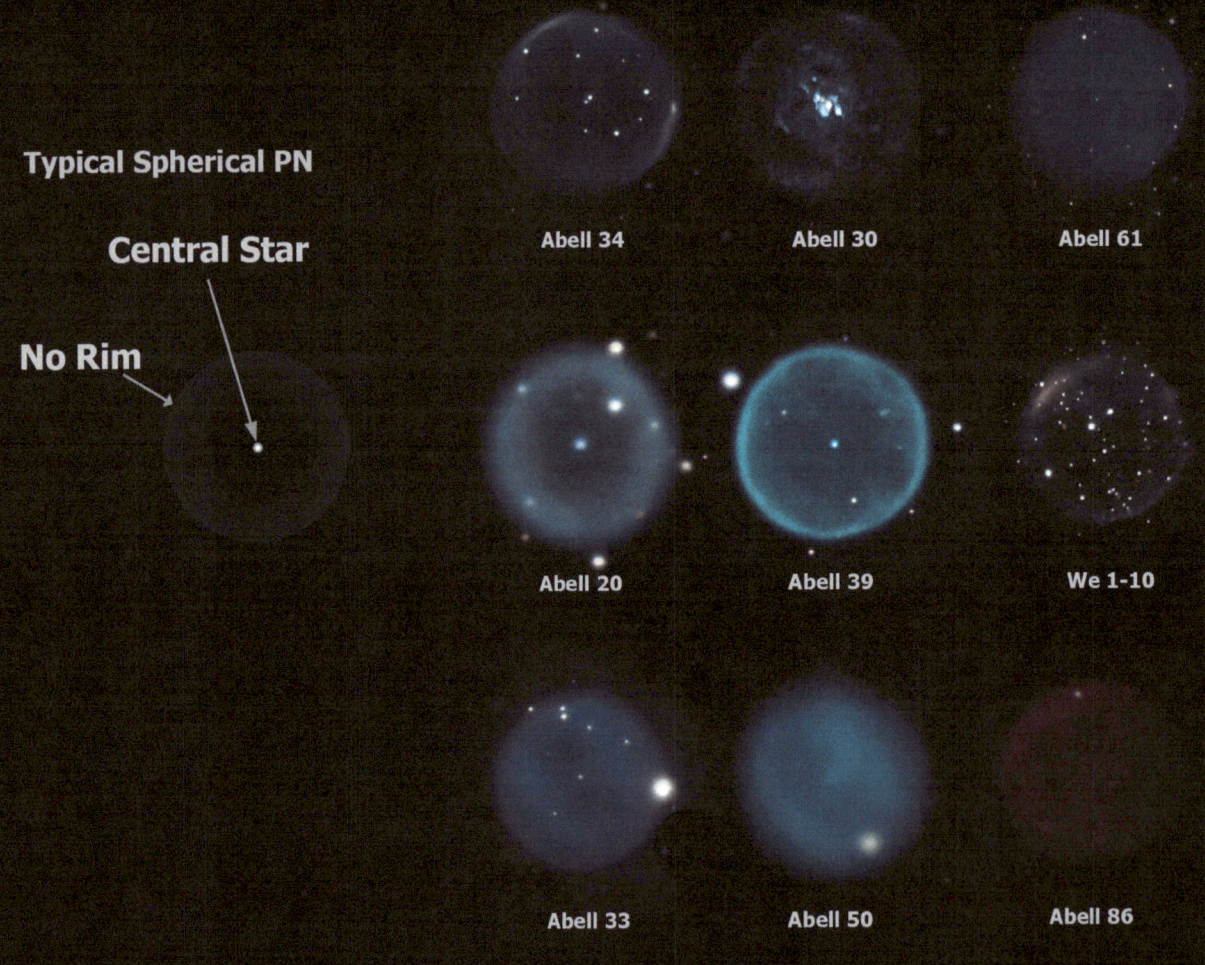

PNe Morphology - Bipolar

About 20% of PNe are believed to be bipolar. These PNe have the most interesting shapes.

To create such an axisymmetric and complex system, a collimation mechanism is need. That could be provided in various ways, but it seems that the most likely way is created when the progenitor is a binary star system. One of the stars, late in life during its AGB phase, grows so large that its outer envelope forms a swirling equatorial disk (torus) around the companion. The disk constrains the companion's bi-polar flow, forming 2 polar lobes which expand and may break through over time.

It is important to note here that while the 2 companions are interacting, they avoid the common envelope phase for most of the interaction time (unlike the elliptical PN development described on the next page). The "common envelope phase" is when both companion stars of the binary system share the same ejected gas envelope of the late life AGB star.

We may see 5 different apparent views of the same bipolar system, as shown and described below:
1. The 1st column shows an end-on view, where the torus appears to us as a ring.
2. The 2nd column shows an inclined view, where the torus appears as an ellipse.
3. The 3rd column shows a side-on view, where the torus appears as a rectangle. Here it is easy to see the hourglass lobe shape.
4. The 4th column also shows a side-on view, but this time with lobe breakout.
5. The 5th column shows multipolar examples, where multiple bipolar flows are occurring. It is believed that this is due to changes in the orientation of the binary star system and its polar ejections over time.

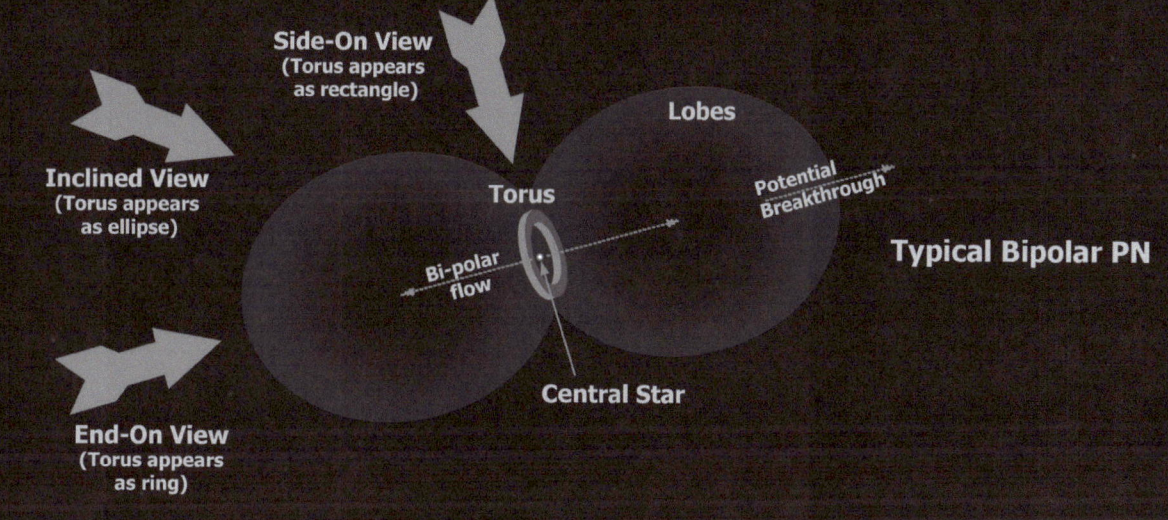

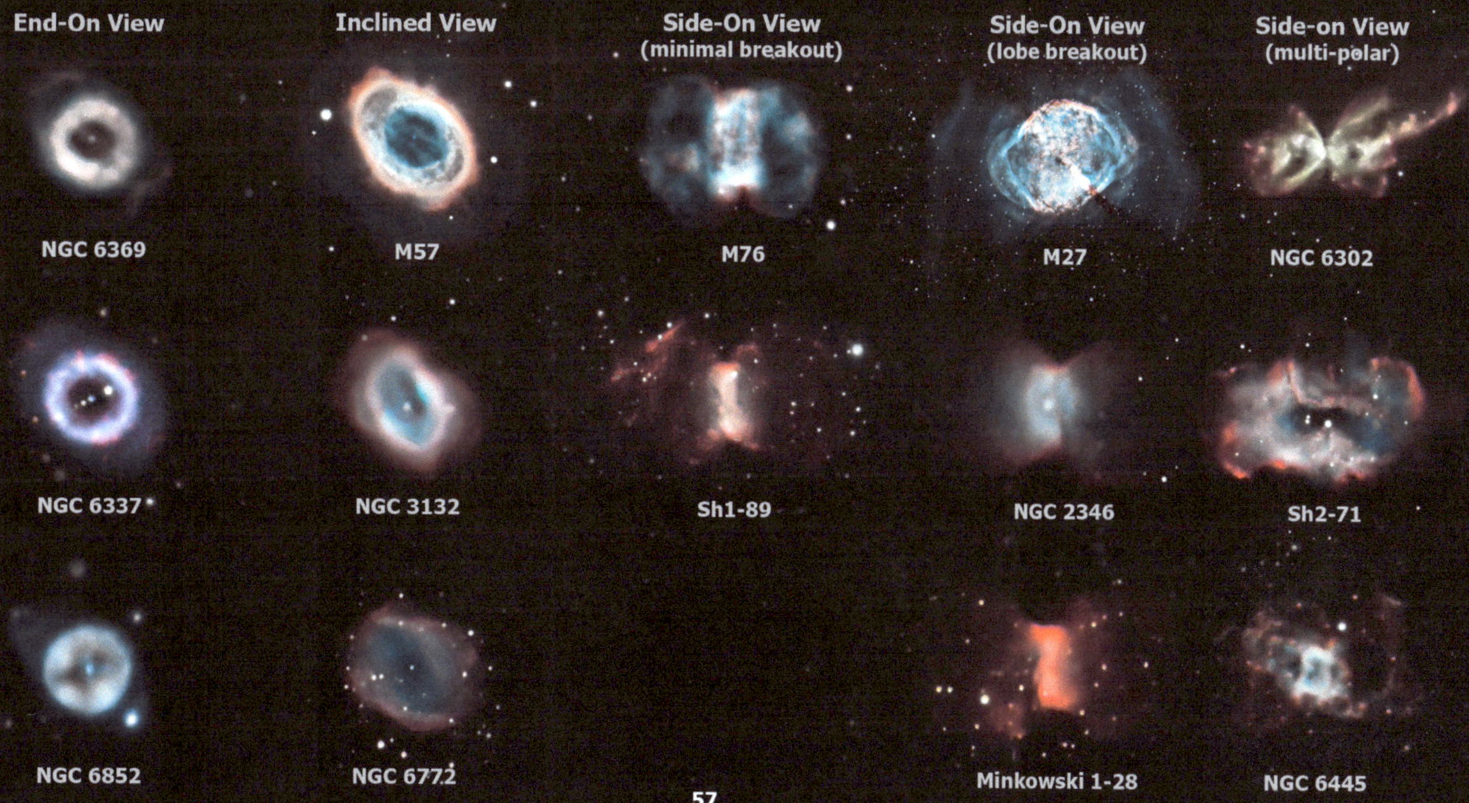

Typical Bipolar PN

End-On View	Inclined View	Side-On View (minimal breakout)	Side-On View (lobe breakout)	Side-on View (multi-polar)
NGC 6369	M57	M76	M27	NGC 6302
NGC 6337	NGC 3132	Sh1-89	NGC 2346	Sh2-71
NGC 6852	NGC 6772		Minkowski 1-28	NGC 6445

PNe Morphology - Elliptical

The majority of PNe are believed to be elliptical. Scientists believe that there may be 3 underlying reasons for an elliptical pN shape. The first 2 result in a truly elliptical shape as seen in the left half of the page below, while the last is related to our viewing angle of a bipolar nebula and is seen on the right half of the page below.

The first reason for an elliptical shape is the nature of the progenitor star. Like the bipolar case are the previous page, the progenitors of these elliptical PNe are believed to be binary star systems. For this elliptical case, however, the companion star orbits closer to the progenitor star so that it lies within its envelope for at least a portion of the formation time. The resulting nebula then takes on an elliptical shape.

The second reason for an elliptical shape is a variation of the nature of the winds of the progenitor from the pole to the equator. Higher rotational speeds of these stars could lead to a stronger equatorial wind, leading to a nebula shape that deviates from that of a perfect sphere.

The third reason is seen in the schematic on the right, where a bipolar PN is viewed from an inclined view such that the central bright torus appears to be elliptical. In this case, the resulting elliptical nebula may have a thicker ring and more texture.

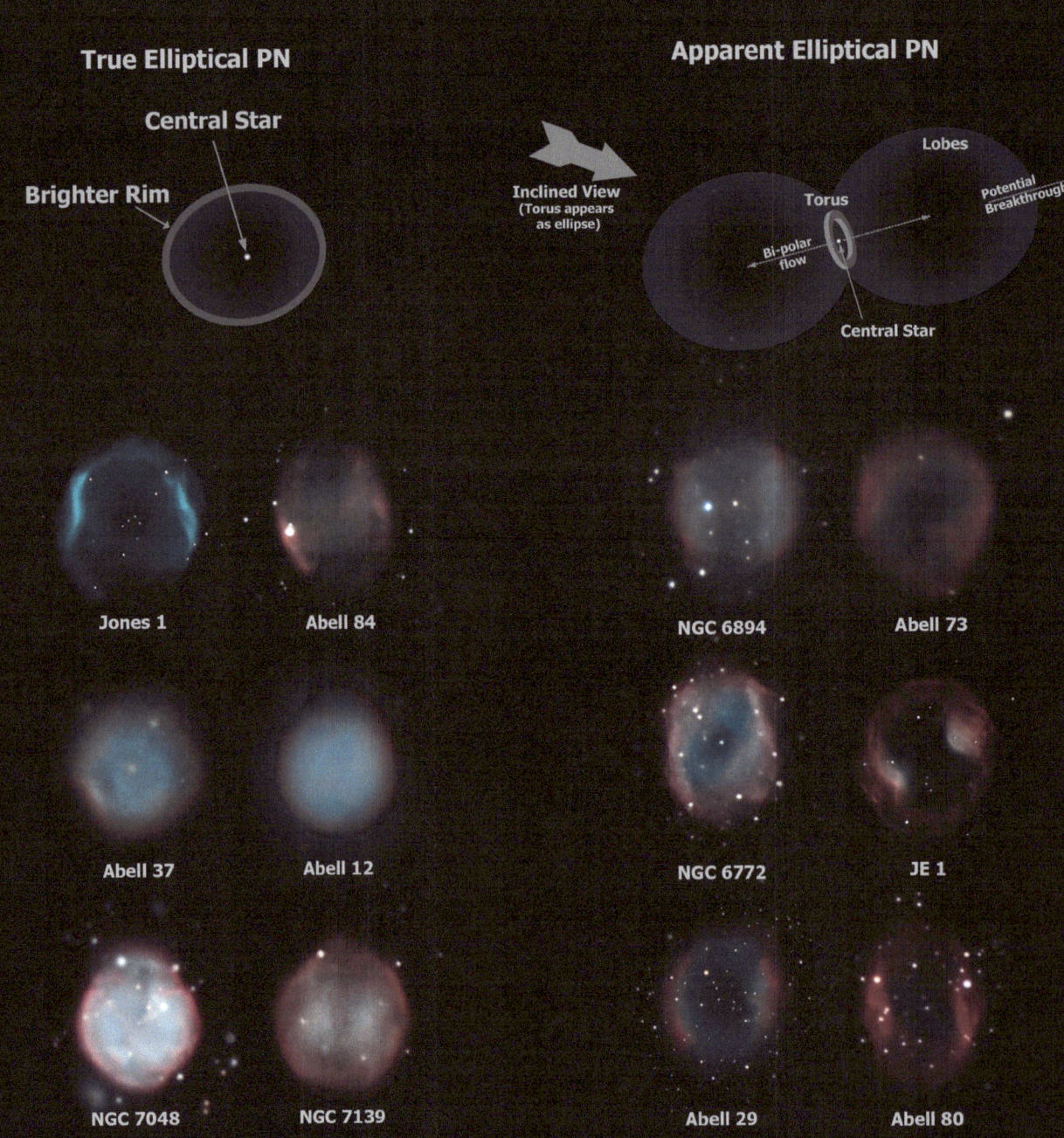

PNe Morphology - Peculiar

The complexities of the PN formation process leads to many unusual PNe with appearances that are difficult to understand, much different than the standard spherical, bipolar, and elliptical structures seen on the previous pages. These complexities may due to a variety of factors, often in combination, some of which are listed below:

1. Multi-polar evolution: As stated above, often the binary star systems axis alignment changes over time, resulting in multiple directions of outflow and a more confusing (but more beautiful) multi-polar shape.
2. Binary star configuration: The orbital separation, speed, and mass of the progenitor pair influences the PN shape. Over time, these variables can change, leading to different PN shapes over time.
3. ISM interaction: Interaction of the gas and jet PN flows with the interstellar medium will often lead to asymmetric PN behavior.
4. Local gravitational effects: The presence and influences of large planets in the progenitor's solar system could lead to disturbance in the interior of the nebula.
5. Magnetic fields: Magnetic fields, like those of our sun, are believed to have the potential to influence the PN shape.

It is also possible that some of these objects are not actually PNE but some other form of emission nebulae. Massive Wolf Rayet stars in particular can create unusual structures. Distant HII regions also have been mistaken for PNe.

Some of my favorite peculiar PNe are shown below:

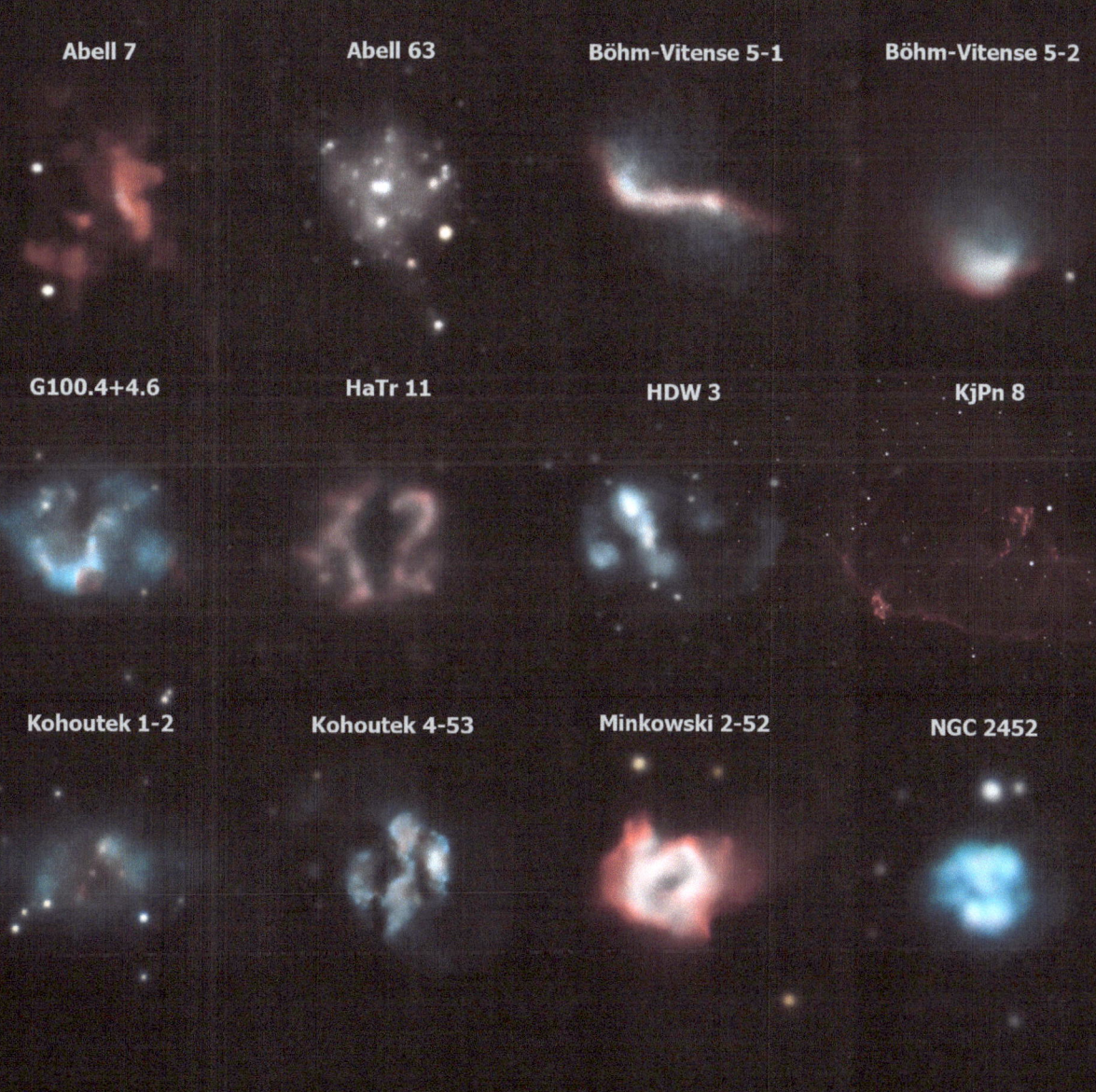

PNe Classics

PNe come in all shapes, structures, colors and sizes. Some of the them are classics, resembling familiar shapes that we have seen elsewhere in the sky. That should not be a surprise, given the number of PNe in the sky and the identical physics at the root of their behavior. On this and the following 4 pages, I present 5 classic nebulae types - Ancient, Diamond Ring, Owl, Ring and Little Dumbbell.

PNe Classics - Ancient

What is an ancient planetary nebula? They are typically identified as a large, faint planetary nebula. But the scientific definition is not as clear as one might expect. Most agree that the term refers to a large, faint planetary nebula. The logical definition of "large" would be actual size, but numerous papers equate large in this context to apparent size, which seems odd to me. That may be due to the difficulty of determining distance for these faint objects.

The average actual diameter of these PNe, where that number can be determined, is about 5 light years. In my experience, the average PN diameter is about 1 light year. All of the PNe in the poster are greater than 6 arc-minutes in apparent diameter.

I have not included here objects which have been considered to be ancient PNe by some sources, but which are not now believed to be PNe. Examples include Abell 35 (a stromgren sphere) and Sh2-176 (a HII region).

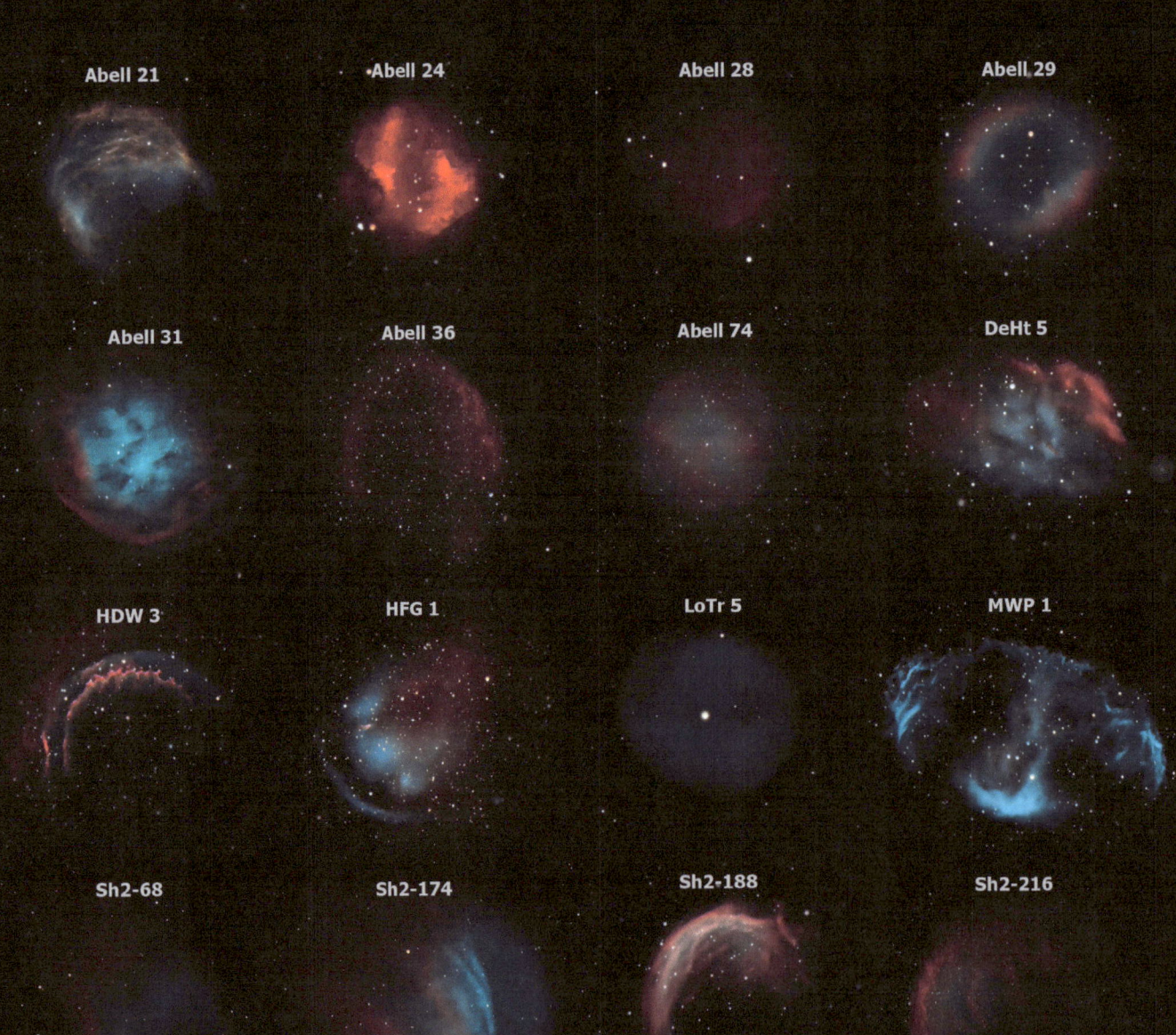

PNe Classics - Diamond Ring

Abell 33 is a planetary nebula which is circular in form and which has a bright star superimposed on the rim. It is a beautiful object and it is easy to see why it has earned the nickname of the Diamond Ring. The star placement on the rim is perfect and the PN is almost a perfect circle.

Although Abell 33 is the classic and cannot quite be matched, other planetary nebulae also have some resemblance to a diamond ring shape. The poster contains a total of 11 of these objects.

Normally I post my images north up, but on this page I rotated all of the images so that the "diamond" is up.

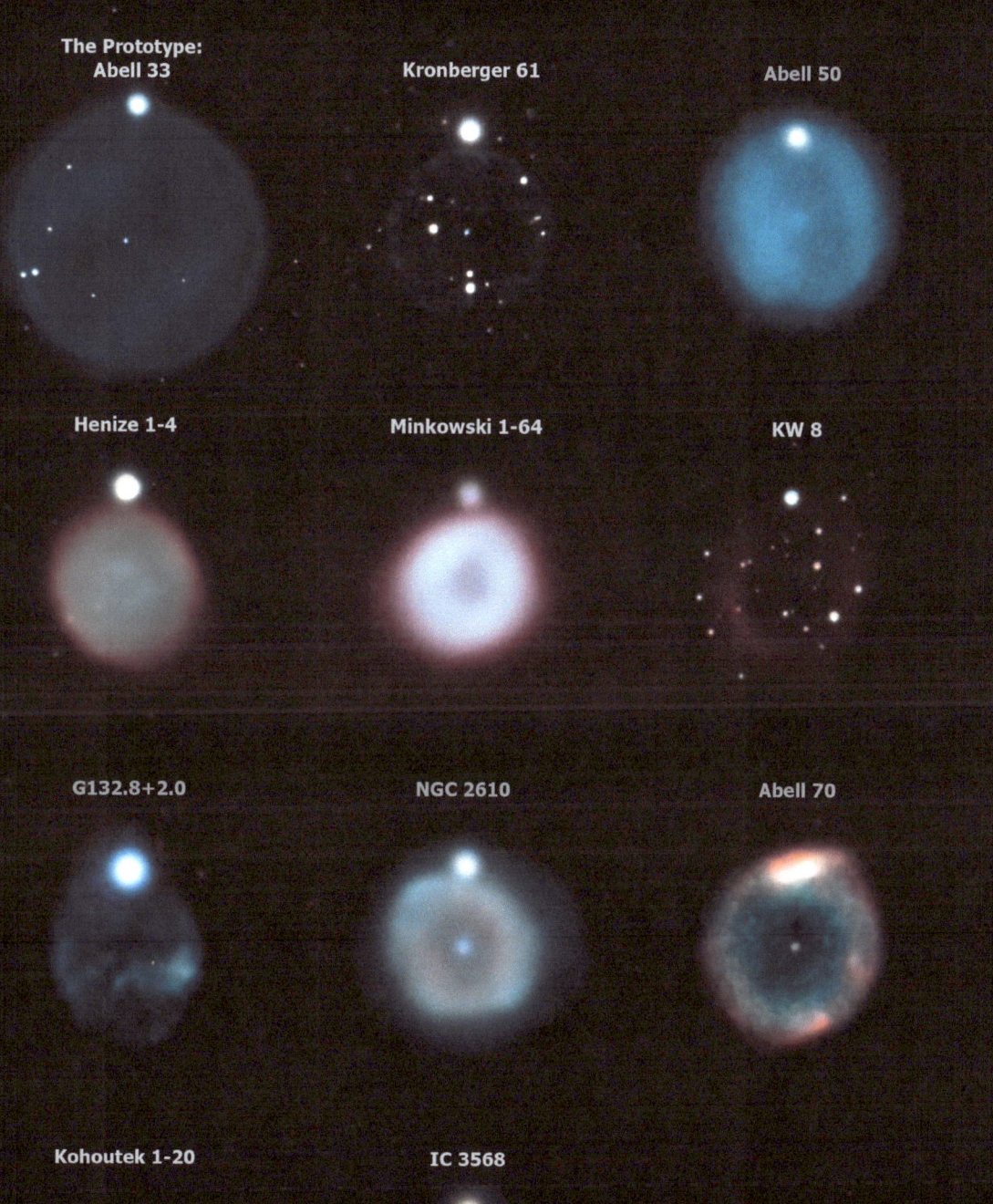

PNe Classics - Owl

The fantastic Owl Nebula (M97) may seem to be a unique object, but a number of other PNe have a similar look to the Owl (although in less detail, since they are located further away). This poster shows 6 PNe which I view as "Owl-type" planetary nebulae, similar in structure and formation to M97.

At first, these nebulae appear to be bi-polar, like most pN. But examining their disks carefully, it seems likely that the Owl-type PNe have spherical shapes - they are almost exactly circular in shape and uniform in brightness, with no strong brightening near the rim. The dark void areas are what trick us, in terms of creating a "bi-polar" look. Scientists still do not understand the cause of the bi-polar look to the dark voids.

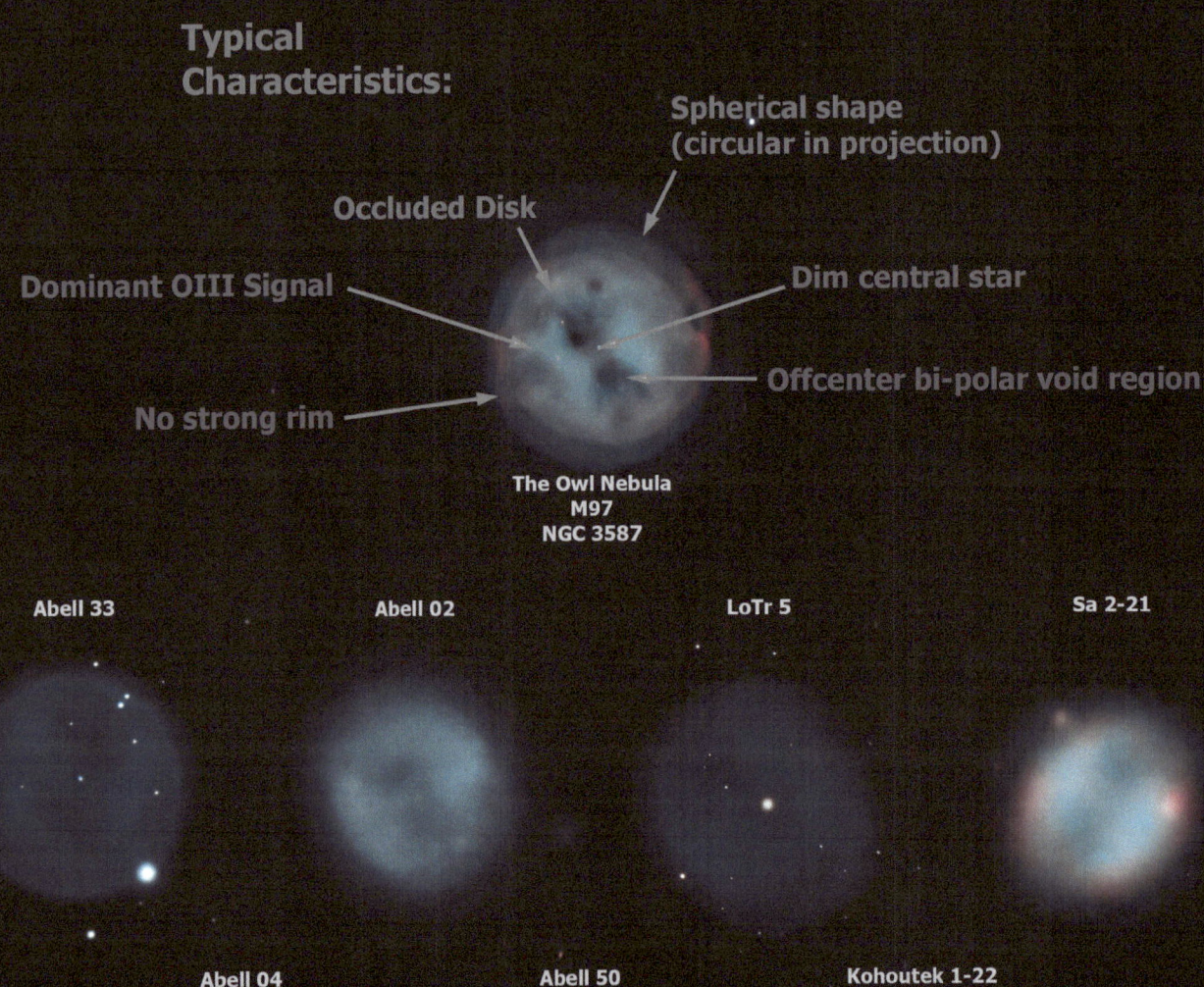

PNe Classics - Ring

There are many ring-shaped planetary nebulae in the sky but only one Ring Nebula. One of the most striking objects to see visually by telescope, the Ring Nebula is unusually bright. In fact, it is one of the few planetary nebulae that I had to image narrowband at 1 minute subs (instead of 5 minute) to avoid oversaturating.

This nebula is a bi-polar nebula and our view is aligned at about 30 degrees from the bi-polar axis. Around the center of the nebula, a torus of dense dust creates the bright white region by constraining the bi-polar outflow into a dense ring, which we see at a slight angle that presents an apparent oval shape instead of a circle. In addition to constraining and concentrating the gas, the torus directs the outflow into two large opposing end-on lobes, which are the faint outer regions that we see in the image.

Nothing else looks exactly like the Ring, but some of the other PNe below seem to have a similar structure.

M57
(Ring Nebula)

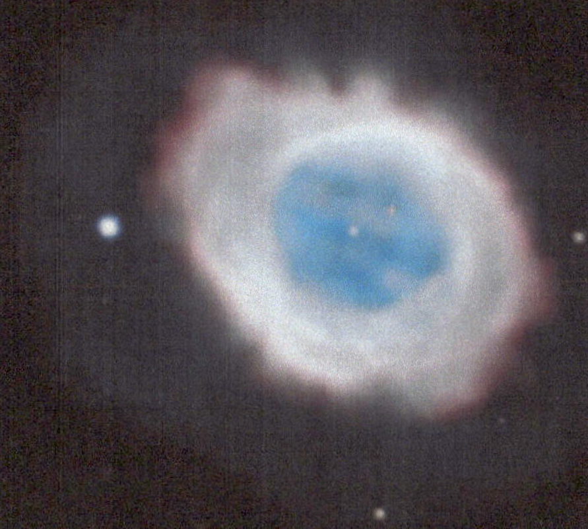

NGC 6563

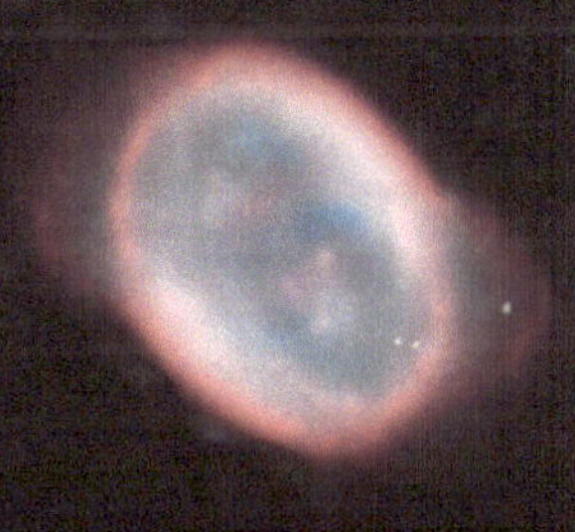

NGC 6772

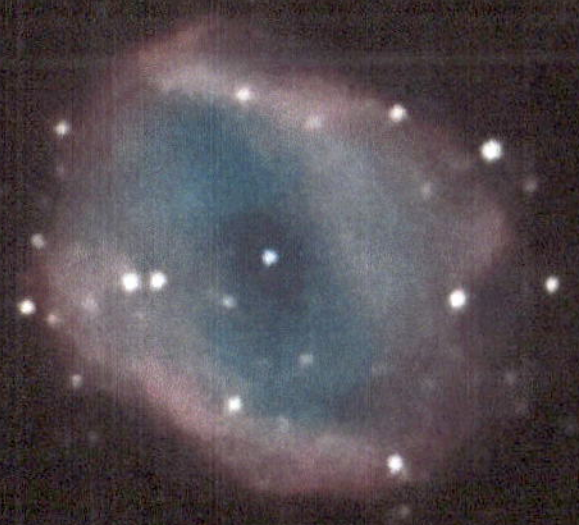

NGC 7293

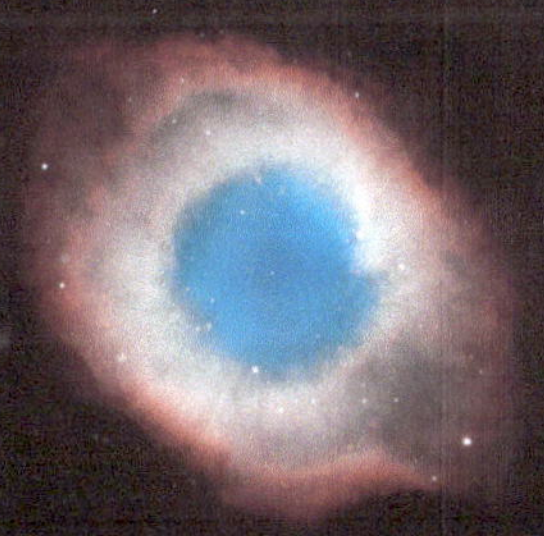

NGC 3132
(Southern Ring)

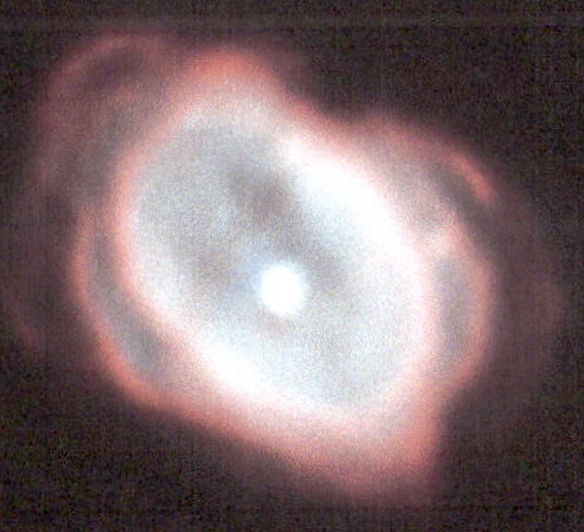

PNe Classics - Little Dumbbell

This is famous planetary nebula M76, nicknamed the Little Dumbbell Nebula. This magnitude 17.5 object is the dimmest nebula in the entire Messier catalog. The Little Dumbbell Nebula is named after its visual appearance in a telescope, where you can see only a very faint, grey dumbbell shape of the central bright region. Photographic images of this object more resemble its alternative nickname, the Apple Core Nebula. This planetary nebula is clearly bi-polar, with some OIII breakout at each end.

I have included several other PNe below where we have a similar perspective view, looking side-on at a bipolar nebula so that the bright central region takes on a rectangular shape instead of a donut shape.

I find it interesting that all of the other PNe on this page are HII-dominated, unlike M76 which is OIII-dominated.

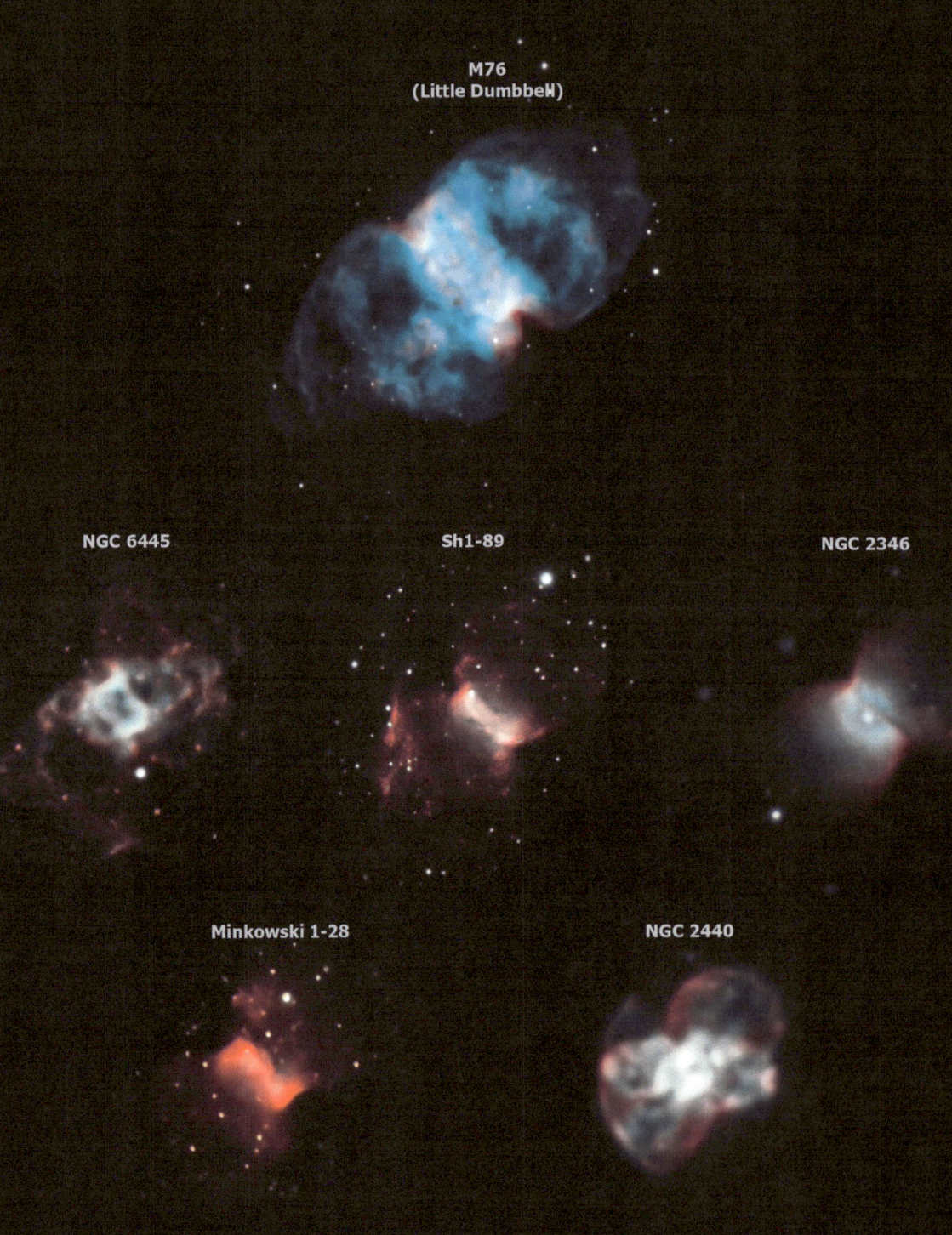

PNe Features

So far, the subject topics have been PNe overall structure and type. We now focus on specific features of PNe. This page, and the 7 that follow, focus on 8 interesting features that are unique to PNe objects in the universe.

PNe Features - Ansae (FLIERs)

Ansae are much simpler to describe than understand. They appear as a pair of tiny red objects found left and right of OIII-dominated elliptical planetary nebulae, when the major axis of the PN is oriented horizontally. The ansae often appear to be leaving a faint "jet" trail, connected them with the progenitor star and appearing to be bipolar ejections.

I have turned all of the PNe below from their north up orientation so that the major axis is horizontal. The Latin word ansae means "handle", and comes from the field of archeology where it particularly describes the looped handle of a ancient vase.

Studies of ansae reveal that they are moving much faster than the surrounding gas, expanding faster than the nebulae itself. They also are regions of low ionization, showing up brighter in image taken through low ionization filters than through those showing higher ionization (like OIII). For these reasons, ansae are also described as FLIERS, which stands for Fast Low Ionization Emission Regions.

FLIERS could be an early stage of BRETs, described on page 67.

NGC 7009   Minkowski 3-1   IC 2149

IC 4634   NGC 6826   Abell 7

PNe Features - Jets

PNe ansae (described on the previous page) often appear to be leaving a faint "jet" trail, connected them with the progenitor star. They seem to be bipolar ejections and are often described as high velocity, collimated outflows.

The objects below have longer jet trails than on the previous page with the end points no longer being red. They appear to be a snapshot in time later than those on the previous page. The jet trails are longer, extending beyond the main nebulae, and become curved in a point symmetric manner.

NGC 6301   NGC 4361   NGC 40

NGC 6309   ETHOS 1

PNe Features - BRETs

BRETs are subtly different from the Ansae and Jet features described on the previous 2 pages. BRETs stands for Bipolar, Rotating, Episodic jeTs. While the previous 2 features appear to be steadily bipolar, these PNe have signs that the jet ejections have occurred at different times and in different directions, perhaps by a binary star whose interactions are causing fluctuations in the ejection orientations. The disturbance in each nebulae created by these BRETs is incredible.

The word "confirmed" appears below for those objects with supporting kinematic data that the deformations are high velocity jets.

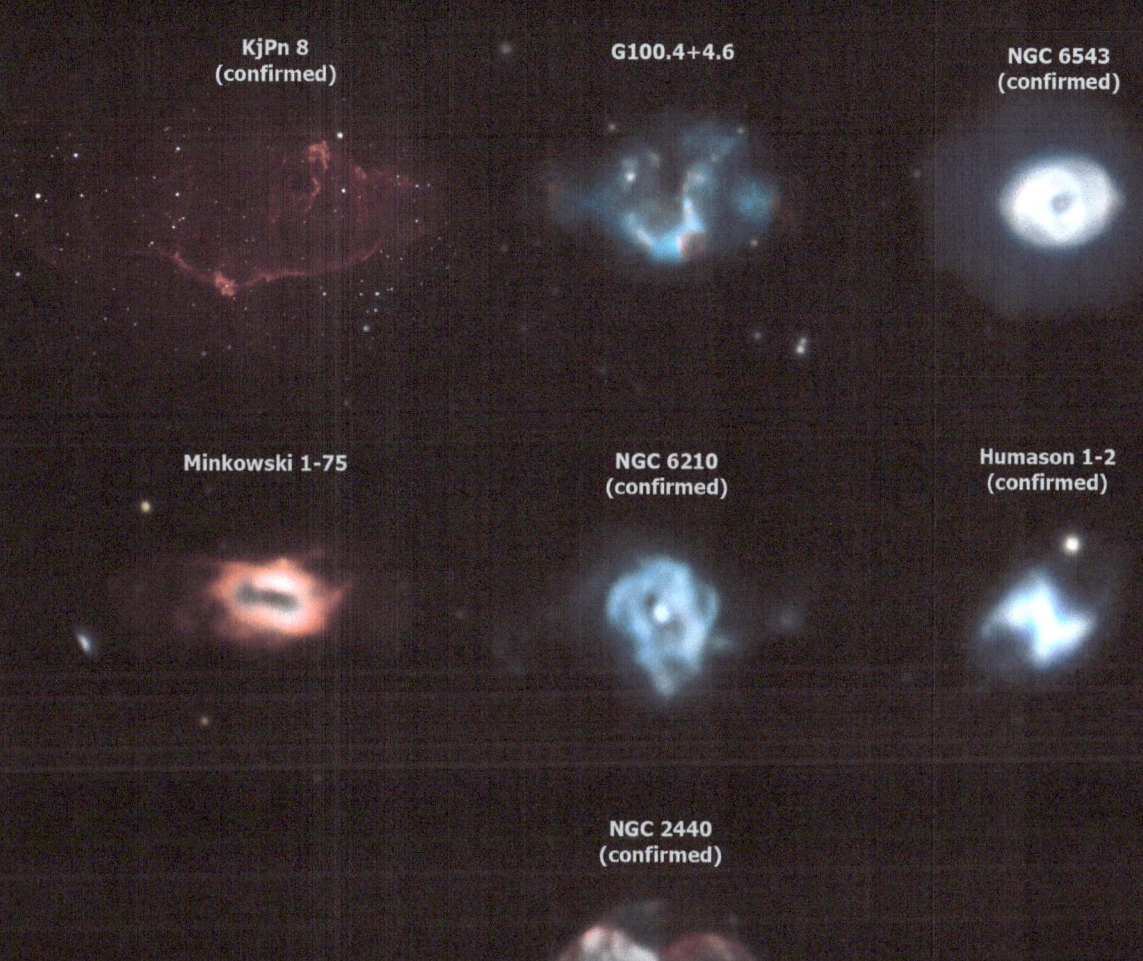

PNe Features - Filaments

Some PNe show signs of filamentary structure. These microstructures usually seem to occur around the progenitor star. The filaments appear similar to what is seen in supernova remnant (SNR) nebulae although at a much larger scale.

Scientists are still working to identify the cause of this fascinating structure. The progenitor star is bright in each of these PNe. The lack of any HII-dominated PNe with filaments is striking.

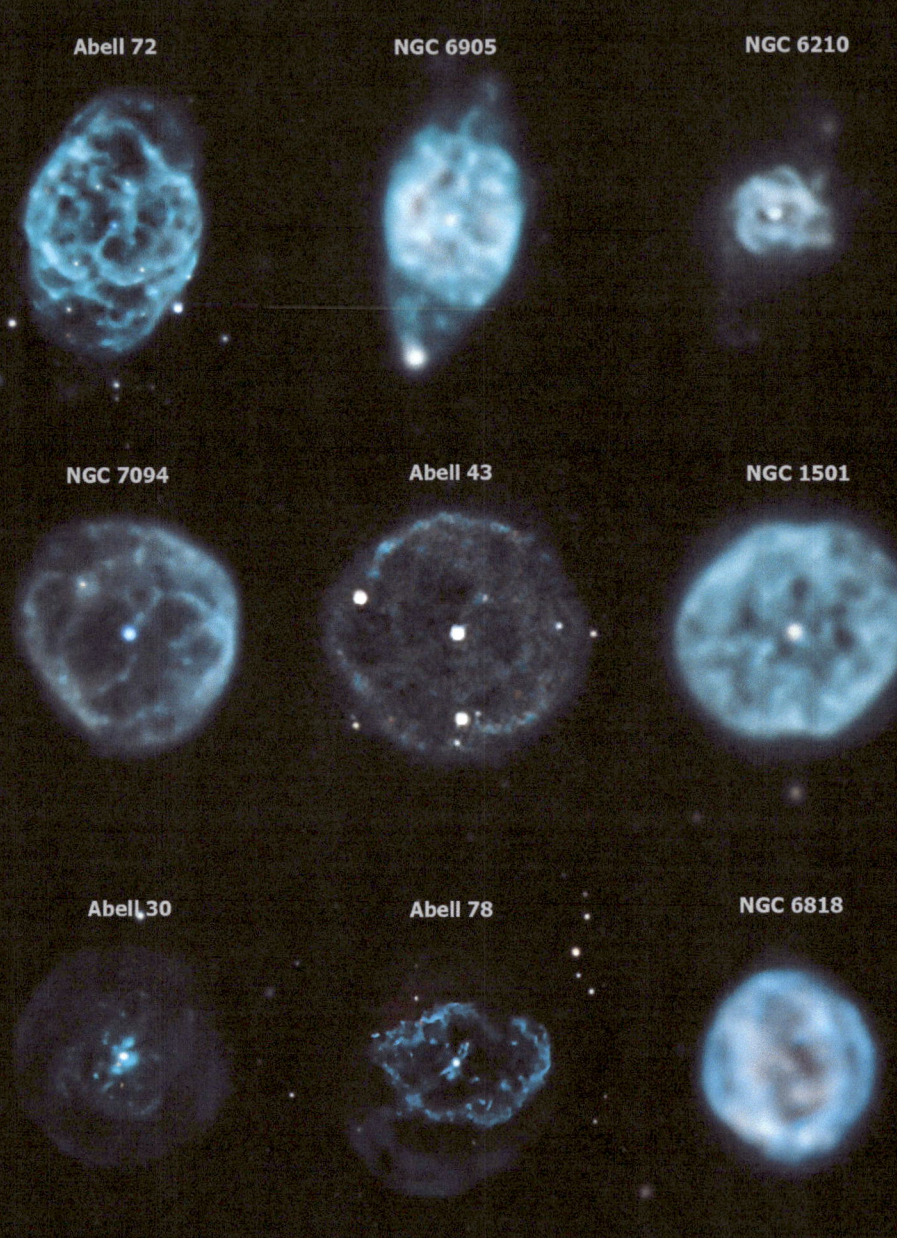

PNe Features - Textures

The interiors of most PNe are transparent or semi-transparent, especially at the center. Not so with the nebulae on this page.

The classic PNe texture feature is seen on NGC 246 below, nicknamed the Skull Nebula. Texture can be thought of as an occluded nebula with signs of broad filaments and dust. As has been seen before in other pages of this book, the only PNe with this feature are OIII-dominated.

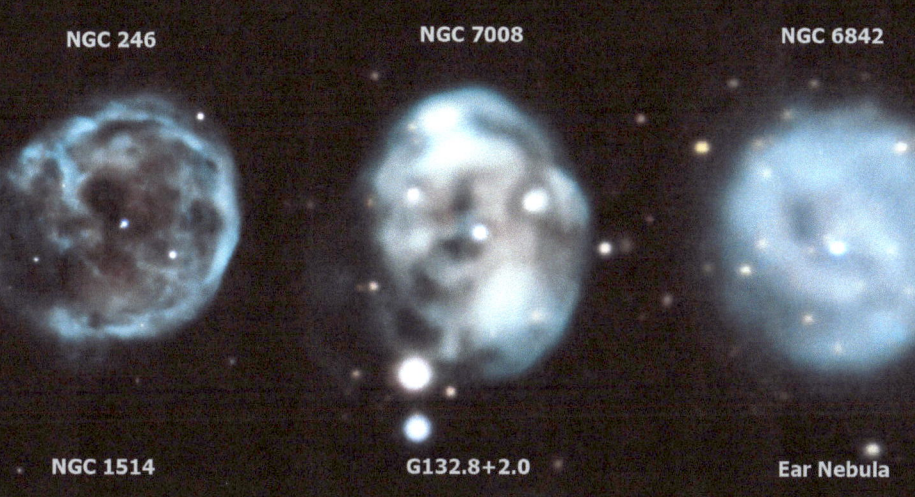

PNe Features - End-on Torus

As shown on Page 57, bipolar PNE are one of the 3 main PNe types. In a bipolar PN, a torus of gas and dust accumulates around one of the binary progenitor stars, created from the outer envelope of the companion late life AGB star. When viewed from the end, as depicted below, this constricting torus appears as a bright thick ring. This ring often appears white due to the balance of OIII and HII signal in this region.

Each torus seen in the images below is circular, indicating that our view of these PNe is near end-on. Many toruses (most notably M57) appear elliptical, indicating that they are inclined to our line of sight.

In some cases below only the torus is visible. It is believed that in these PNe the extended bipolar lobes are too dim to be seen.

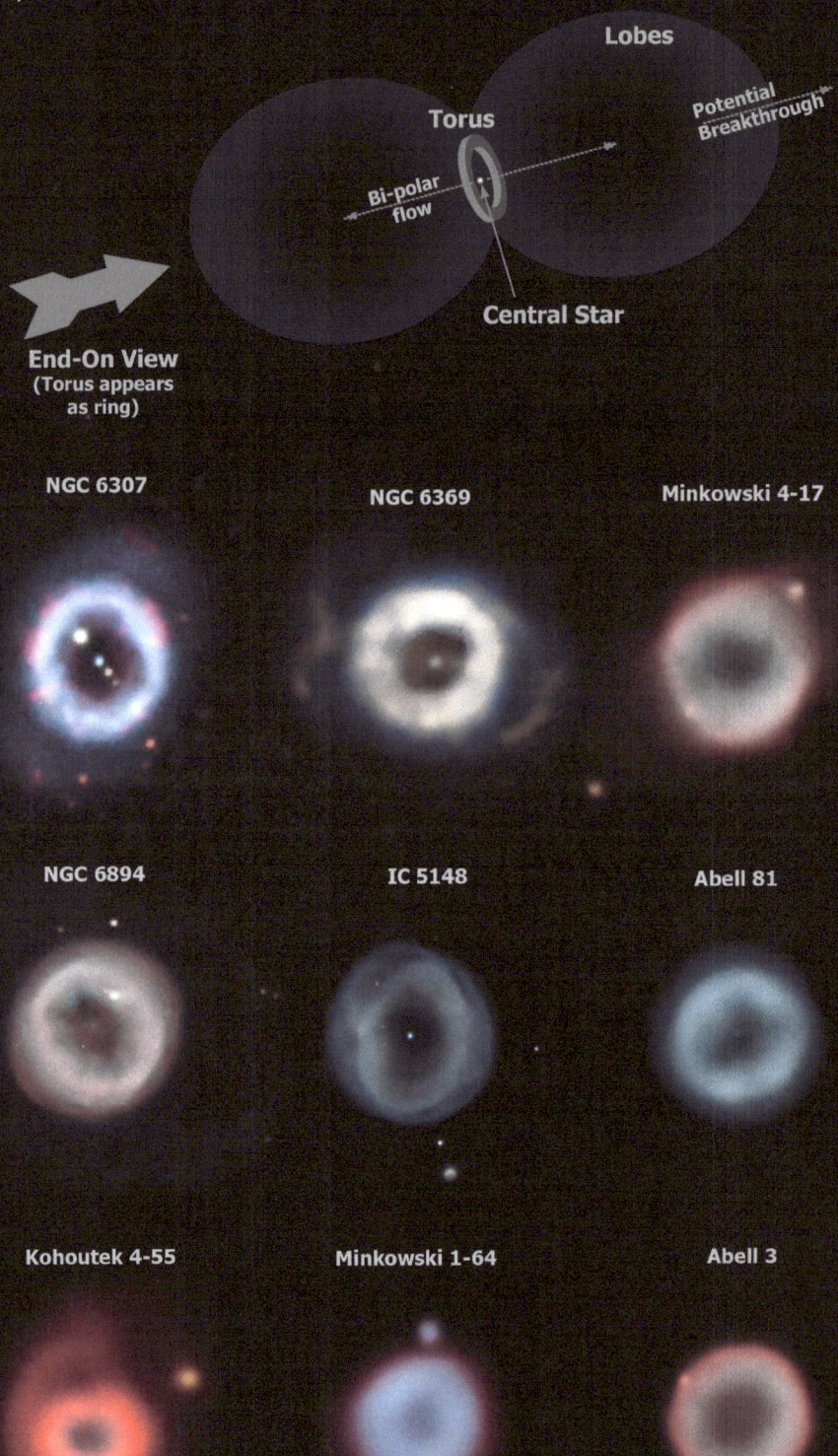

NGC 6307 NGC 6369 Minkowski 4-17

NGC 6894 IC 5148 Abell 81

Kohoutek 4-55 Minkowski 1-64 Abell 3

PNe Features - Irregular Torus

Unlike the wide circular toruses of the previous page, these toruses appear thinner and more irregular. Also, there are no bipolar lobe extensions to be seen. These PNe may be younger in age and still expanding.

As seen in many of the previous pages, all of these PNe are OIII-dominated.

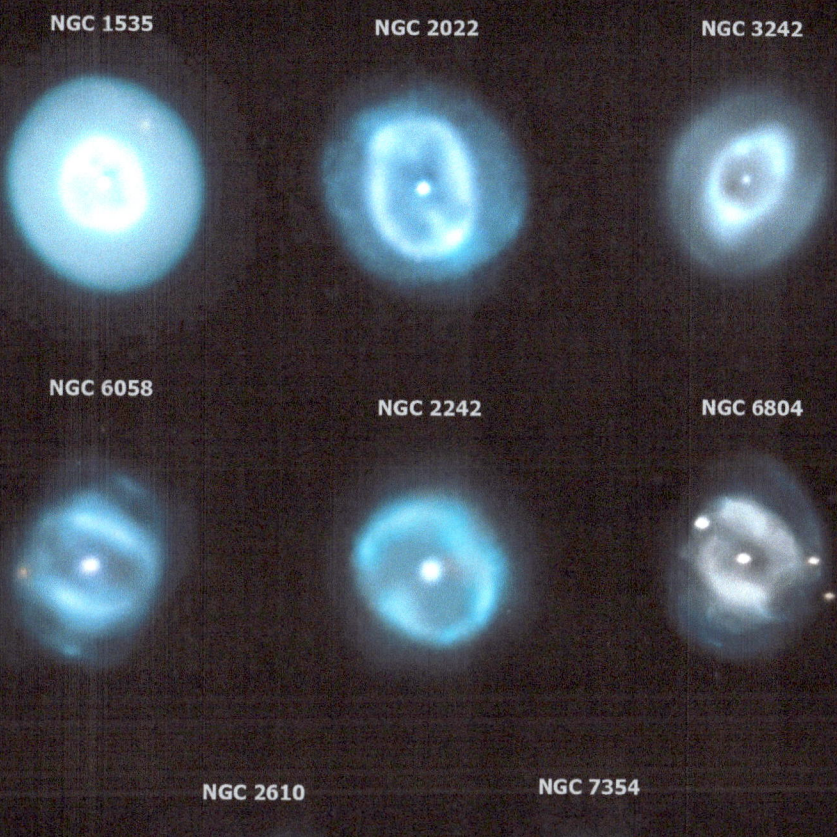

PNe Features - Rotating Elliptical

Many of the PNe features have been dominated by OIII signal.   Not so with these PNe. These are elliptical shaped PNe where the outer lobe rim seems to be slightly changing position with time. In all of them, a faint OIII signal extends towards the inner portion of the nebulae and a progenitor star is faintly seen.

The shapes here could be indicative of a binary progenitor star system where the orientation of either rotation or polar ejection changes with time.

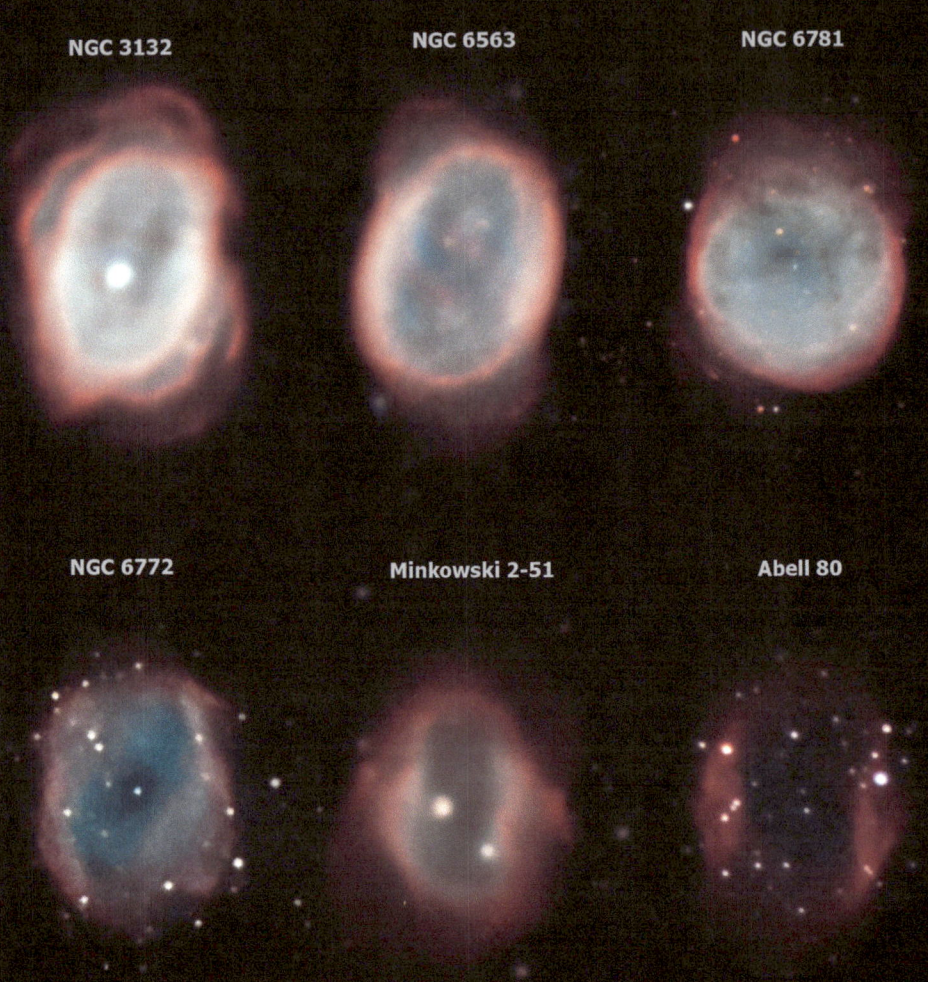

Catalogs

PNe show up in both general catalogs and in PNe-specific catalogs. The following pages shows the PNe in 6 different catalogs.

Catalogs - Messier

The Messier Catalogue is the most popular catalogue in astronomy. In the middle of the 18th century, during the return of Halley's comet, a French astronomer named Charles Messier began a life-long search for comets. He would eventually discover 15 of them. For his own convenience and to avoid confusion, he began keeping a journal of sky objects such as galaxies, nebulae, and star clusters which could mistakenly appear as a comet through a small telescope. Over time, this journal evolved into the Messier Catalogue, a list of 110 of the most famous deep sky objects above -35 degrees declination.

The 4 PNe of the Messier catalog are shown below. The PNe are shown to scale:

M27 (Dumbbell)

M57 (Ring)

M76 (Little Dumbbell)

M97 (Owl)

Catalogs - Caldwell

The Caldwell Catalogue is a list of 109 sky objects published by Patrick Moore in 1995 as a complement to the Messier Catalogue. The Messier Catalogue was developed as a list of objects to avoid when looking for comets, not as a list of the best objects in the sky to observe. Messier did not include many of the sky's brightest deep-sky objects, and only included objects he could see from Paris. The Caldwell Catalogue covers the entire sky and includes interesting objects which are not on Messier's list. Both of these lists were compiled for visual use, but both are also convenient lists for astrophotographers.

The Caldwell objects are numbered from 1 to 109 in order from northernmost to southernmost objects, spanning from +85 to -80 degrees declination. Of course, unless you live near the equator, you are not able to see all of the Caldwell objects from one site. The Caldwell catalog includes 13 PNe, but I can only reach 11 Caldwell PNe from my backyard in East Texas. As with the previous Messier page, the Caldwell PNe below are to scale.

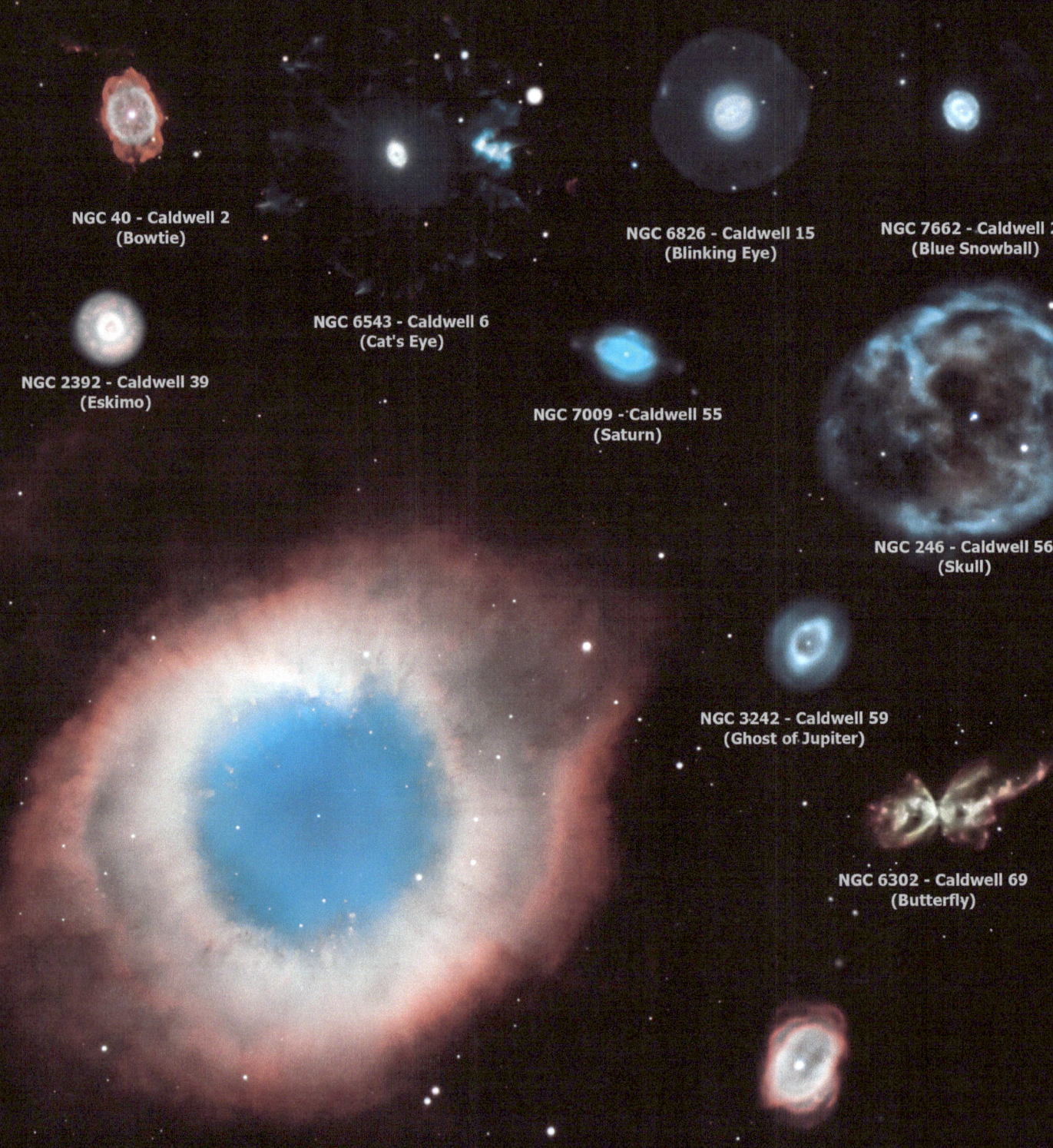

Catalogs - Minkowski

One of the more interesting PN catalogs is the Minkowski Catalog. During his work in the 1940s at the Mount Wilson Observatory, Dr. Rudolph Minkowski published 3 papers which doubled the number of planetary nebulae known at that time. Interestingly, Dr. Minkowski did not number these objects. Instead, in later years, the PNe listed in these three papers were recognized by the designators M 1-xx, M 2-xx, and M 3-xx. Later, in 1965, Perek and Kohoutek added additional objects from Minkowski's observations which became known as the M 4-xx list.

The Minkowski catalog contains mostly PNe (207), but also includes large nebula (40) such as the Pelican and Fishhead Nebulae, 1 galaxy, 1 supernova remnant, and 7 stars. Some of the objects also have NGC, IC, and SH2 designations. Many of the Minkowski PNe are stellar in size and not particularly interesting, but there are some good ones.

On this page, I show what I consider to be the best PNe of each of the 4 Minkowski lists. So many interesting PNe are contained here, including:

• The triple star progenitor system of M 1-90 (Sh2-71)
• The Texas-shaped M 2-52
• The point symmetric PN M 3-1, with a beautiful red ansae at each polar end.

As a bonus, I have included 3 of the tiny Minkowski protoplanetary nebula (PPN) at the bottom of the page. PPN are believed to be younger stars that have not yet reached PN status.

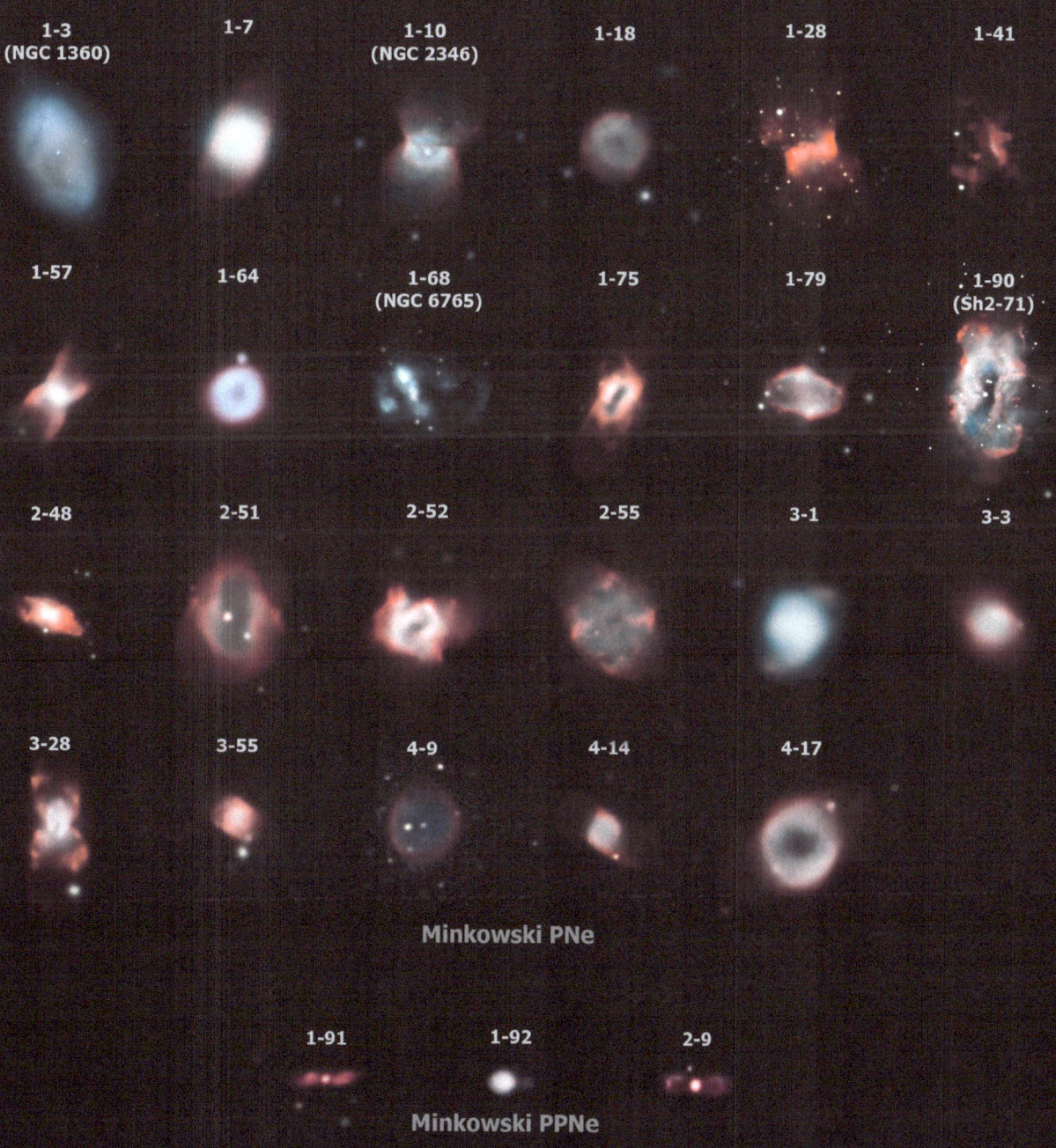

PNe Catalogs - Kohoutek

The PNe of this poster were discovered by the Czech astronomer Dr. Luboš Kohoutek. Dr. Kohoutek discovered over 300 new PNe over his lifetime, many of which now bear his name. The 1500 objects in this catalogue were numbered years after the objects were identified in publications, so the catalogue consists of the typical K 1-xx, K 2-xx, and K 3-xx designations (up to K 6-xx) corresponding to their discovery publications.

Dr. Kohoutek's most famous discovery was the comet bearing his name. As Wikipedia states, "Because Comet Kohoutek fell far short of expectations, its name became synonymous with spectacular disappointment." From my perspective, Dr. Kohoutek will always be remembered as a highly respected and accomplished astronomer.

Like the Minkowski PN catalogue, the Kohoutek catalogue contains many fairly mundane and tiny stellar PNe. But there are also some fascinating objects. I have chosen a selection of 18 PNe for the poster which I found to be the most interesting. Also, like the Minkowski catalogue, you will notice that the larger, brighter and more interesting objects are generally from the first (1-xx) published series of objects. As the list progresses to 2-xx, 3-xx, and beyond, the objects become smaller, more faint and more challenging to image.

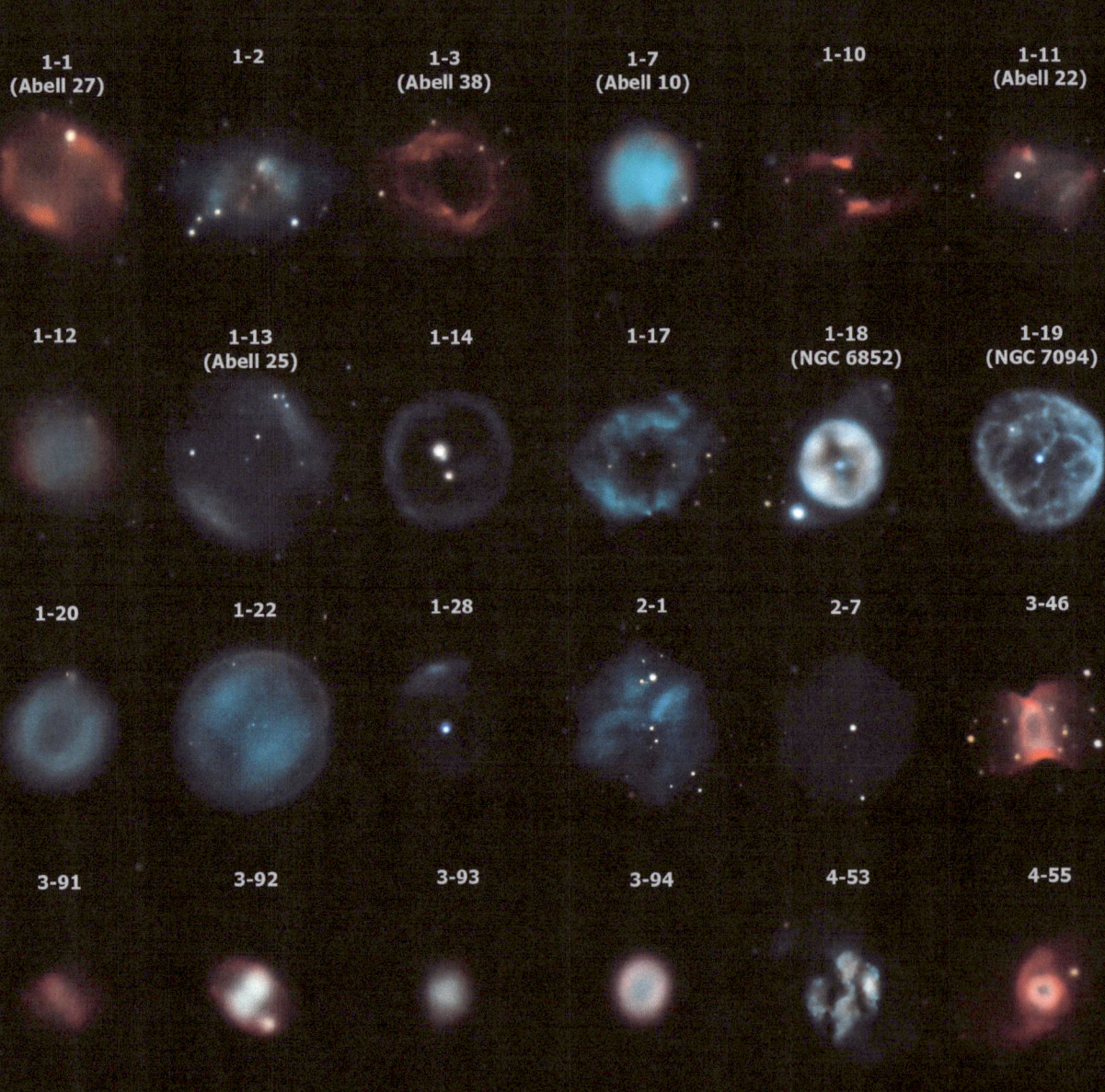

PNe Catalogs - Weinberger

Dr. Ronald Weinberger is a Austrian astronomer and writer who has discovered over 100 planetary nebulae and 200 galaxies in the 1970s and 1980s. As his discoveries came later than most of the other PNe catalogs, such as Kohoutek and Minkowski, Dr. Weinberger's discoveries are mostly faint and obscure. Most are also HII-dominated. Many of his discoveries he made jointly with different astronomers. His sole discoveries begin with We, while joint discoveries add the initials of the co-discoverers.

Catalogs - Abell

The facing page shows the complete Abell PN Catalogue. This catalogue was created in 1966 by Dr. George Abell from new objects discovered on the photographs taken for the Palomar Sky Survey.

The catalogue consists of 86 total entries, although only 78 of these entries are truly PNe. The 8 objects which have been confirmed NOT to be planetary nebulae are not included on the facing page. These non-PNe include reflection nebulae, plate flaws, and galaxies.

It is interesting to examine these PNe, observing commonalities and differences. The descriptions below are simply my observations and are not based on confirming technical data:

Composition (Color): The catalogue is fairly evenly split into 4 different composition types - OIII (blue), HII (red), Balanced (grey), and Mixed (OIII with a HII rim). Note that none of the objects are Mixed in the opposite sense (HII with an OIII rim). This is likely because hydrogen gas is predominant in the early phases, while in later PN life the progenitor stars are only large enough to fuse helium into oxygen and carbon.

Spherical Shape: Only about 1/3 of the Abell PNe appear to be truly circular. Of these circular objects, some appear to be spherical PNe. These are believed to be the only PNe originating from a single progenitor central star. These often appear as a soap bubble, with a translucent interior and a faint rim. Overall, about 10% of PNe are believed to be spherical, and the percentage here is about the same. Examples on the poster include 6, 8, 30, 33, 34, 39, and 61.

Bi-Polar Shape: Overall, about 20% of PNe are believed to be bipolar. These PNe have the most interesting shapes. To create such an axisymmetric and complex system, it is believed that the progenitor is likely a binary star system. One of the stars, late in its life during its AGB phase, grows so large that its outer envelope forms a swirling equatorial disk (torus) around the companion. The disk constrains the companion's bi-polar flow, forming two polar lobes which expand (and sometimes break through) over time.

Bi-polar PNe viewed side-on appear like an hourglass or "figure-8", like the prototype object M76. These objects have a bright central torus section with the brightest portion being the red HII signal on the left and right edges. The bi-polar lobes extend upwards and downwards. In this poster, the only classic side-on bi-polar object is 22. In 22, the faint bi-polar sections at top and bottom are seen to create the hour-glass shape.

Many times the bi-polar lobes at top and bottom have not expanded, resulting in fainter rims top and bottom but still maintaining an overall oval shape to the nebula. This is seen in 44, 49, and 80.

Much later in the development of a bi-polar PN, the top and bottom bi-polar sections have already broken out and are no longer visible. This is seen in 14, 19, 55, 82 and 84.

Finally, a bi-polar PN viewed end-on appears as a bright thick torus. Examples include 20, 53, and 81.

Elliptical Shape: The majority of PNe are believed to be elliptical. In this case, like the bipolar case above, the progenitor is believed to be a binary star system. For this case, however, the companion star orbits closer to the progenitor star so that it lies within its envelope for at least a portion of the formation time. The resulting nebula then takes on more of an elliptical shape. These are seen as oval shapes (sometimes rings) with uniform brightness around the circumference, as seen in 9, 13, 18, 26, 27, 40, 47, 48, 52, 58, 59, 63, 69, 72, and 75. Our views of these rings are typically at a slight angle, resulting in the oval shape (instead of a circular ring).

Multi-polar Shape: Multi-polar nebulae have lobes along more than 1 primary axis. This is likely because the polar axis direction is changing over time. I believe we are seeing multi-polar nebulae in 24, 38, 47, 78, and 79.

Progenitor Stars: The presence (and absence) of progenitor stars in these PNe are fascinating to me. The prototype faint cyan central progenitor stars are seen in 16, 20, 25, 30, 31, 34, 39, 46, 51, 52, 58, 60, 61, 62, 66, 71, 74 and 75.
Some progenitor stars seem unusually bright – perhaps they are not progenitor stars at all but are simply nearby stars superpositioned over the exact center of the nebula. These brighter central stars include 14, 15, 19, 43, 63, and 78.

Bright Arcs: Some PNe have unusually bright arc features, appearing as a spider web. These include 21, 43, 72, 78 and 79.
ISM: The impact of the interstellar medium (ISM) on the PN shape, creating an asymmetric object, is seen in many of the PNe, including 13, 21, 52, 59, 62 and 75.

Enigmata: Some of the PNe are simply difficult to understand and resemble HII regions more than PNe in appearance. They contain no OIII signal. These include 45 and 71.

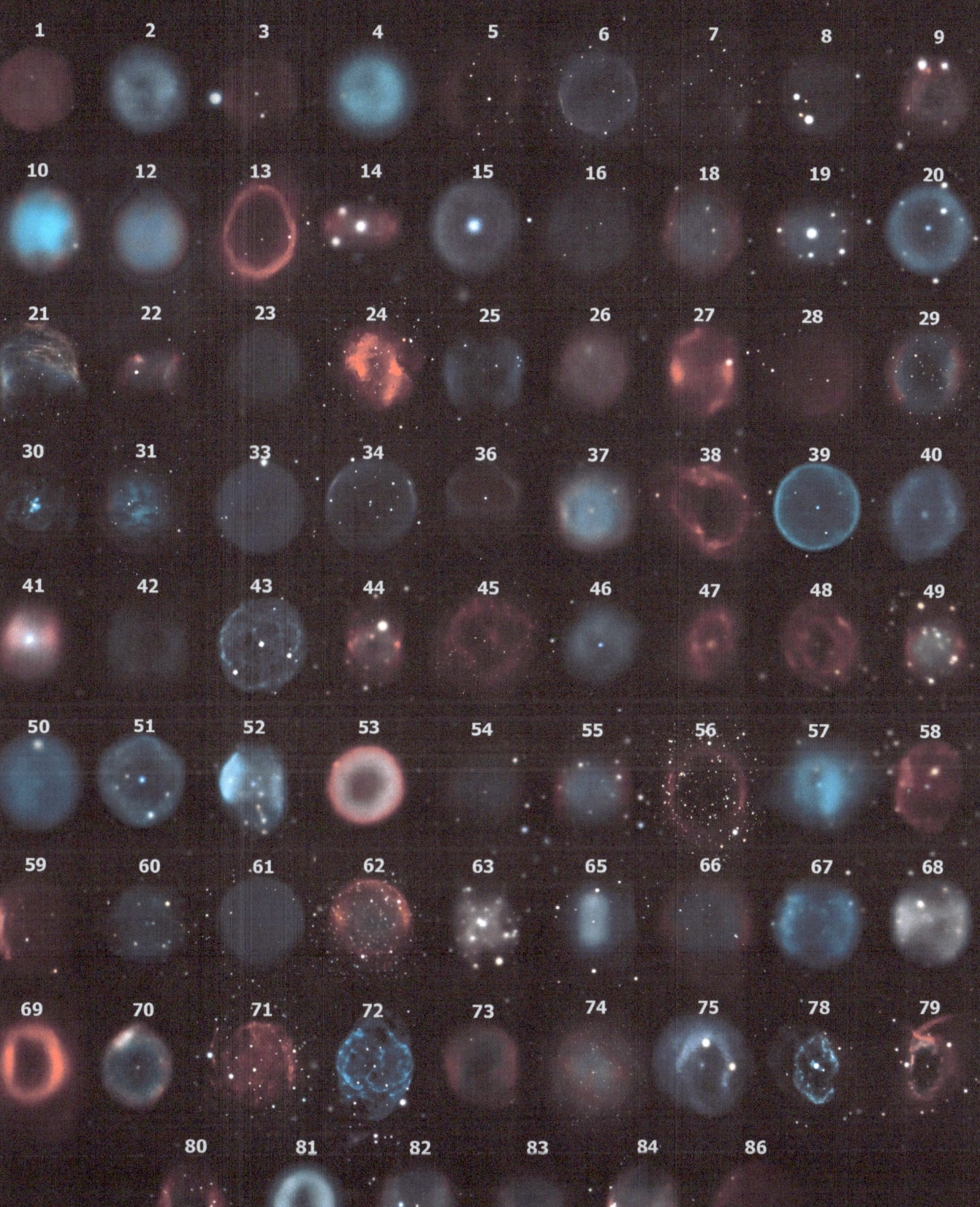

Abell PNe Catalog

Note - 8 non-PN Abell objects not included

Top 250 PNe Highlights

In closing, here is some additional perspective on this collection of 250 PNe to help you decide upon which you should (and should not) target in one of your upcoming imaging sessions.

IC 1295

NGC 6781

Top 10
1. NGC 7293
2. M27
3. M57
4. NGC 6543
5. M97
6. Abell 21
7. M76
8. Jones-Emberson 1
9. NGC 246
10. Sh2-188

Most Underappreciated 10
(shown on this page)
1. Sh2-71
2. NGC 2371
3. Jones 1
4. NGC 40
5. Abell 72
6. NGC 1514
7. NGC 6781
8. IC 1295
9. Sh2-200
10. Sh1-89

Bottom 10
1. IC 2003
2. IC 4593
3. IC 5217
4. Kohoutek 3-93
5. NGC 6741
6. IC 418
7. IC 351
8. IC 3568
9. Kohoutek 3-91
10. Kohoutek 3-94

Sh2-71

Sh1-89

Largest
Apparent Size in arc-minutes
1. 120' - Sh2-216
2. 25' - PFP 1
3. 20' - IsWe 2
4. 20' - IsWe 1
5. 17' - Sh2-174
6. 15' - M27
7. 15' - WeDe 1
8. 15' - Abell 31
9. 13' - MWP 1
10. 13' - Abell 7

Abell 72

Smallest
Apparent Size in arc-minutes
1. 0.1' - IC 2003
2. 0.1' - IC 4593
3. 0.1' - IC 5217
4. 0.15' - NGC 6741
5. 0.2' - Campbell's Star
6. 0.2' - IC 4634
7. 0.2' - Haro 2-8
8. 0.2' - IC 418
9. 0.2' - Minkowski 3-55
10. 0.2' - IC 351

Sh2-200

Jones 1

Brightest
1. 8.3 - NGC 7662
2. 8.8 - NGC 7027
3. 8.9 - LoTr 5
4. 9.3 - IC 418
5. 9.3 - NGC 6818
6. 9.5 - NGC 1514
7. 9.6 - NGC 6826
8. 9.7 - NGC 2392
9. 10.0 - NGC 6629
10. 10.0 - NGC 3132

NGC 1514

Dimmest
1. 21.5 - Böhm-Vitense 5-2
2. 21.0 - Kohoutek 3-91
3. 20.9 - Abell 18
4. 20.0 - Abell 38
5. 20.0 - Abell 42
6. 20.0 - Minkowski 3-55
7. 19.9 - Abell 13
8. 19.6 - Abell 22
9. 19.3 - Abell 71
10. 19.0 - Abell 1

NGC 40

NGC 2371

Furthest North: IC 3568 (+81 deg dec)
Furthest South: IC 4406 (-44 deg dec)

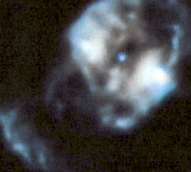

Scholarship

If you would like to help the next generation of astronomers, please consider a small donation to the Imm Astronomy Scholarship Fund, a tax-deductible donation account established to provide financial assistance to college students with an interest in astronomy.   100% of your donation goes to student scholarships. Scholarships are awarded annually each June.

Search for "imm astronomy scholarship fund" on the web to find both the donor site and, if you are a student, the application site.

Final Words

I hope that you have enjoyed seeing the planetary nebulae in this book as much as I have enjoyed imaging and researching them. I particularly hope that you are successful in imaging many of these wonderful objects. There are enough targets here to bring you many years of satisfaction.

Please contact me at immgr@swbell.net if you have any questions, corrections, or suggestions for improvement.

All images by Gary Imm
from Onalaska, TX, USA
(SQM Magnitude 20.8 - Bortle 4.5)

Copyright © 2024 by Gary Imm.
Copyright granted to make copies
for non-commercial purposes.

Alphabetic List of Top 250 PNe (1/5)

#	PN Name	RA (H / M / S)	DEC (D / M / S)	Const.	Transit 9:00 PM	Transit 1:00 AM	Score	Class	Size (')	Distance (ly)	Diam. (ly)	Visual Mag
225	Abell 01	00h 12m 36s	69° 10' 41"	Cep	Nov 13	Sep 14	2	SHN/h	0.80	8200	2.0	19.0
231	Abell 02	00h 45m 36s	57° 57' 24"	Cas	Nov 22	Sep 22	2	SOY/ho	0.50	13000	2.0	15.0
241	Abell 03	02h 12m 12s	64° 09' 05"	Cas	Dec 14	Oct 14	2	SHY/hi	1.2	8400	3.0	16.0
245	Abell 04	02h 45m 26s	42° 32' 36"	Per	Dec 22	Oct 22	2	SOY/o	0.40	18000	2.0	15.0
246	Abell 05	02h 52m 13s	50° 35' 52"	Per	Dec 24	Oct 24	2	EHN/q	2.5	5400	4.0	u
247	Abell 06	02h 58m 53s	64° 29' 59"	Cas	Dec 26	Oct 26	2	SCY/t	3.0	3100	2.7	15.0
14	Abell 07	05h 03m 08s	-15° 36' 13"	Lep	Jan 26	Nov 26	2	AOY/aq	13	1800	6.7	16.0
15	Abell 08	05h 06m 38s	39° 08' 08"	Aur	Jan 27	Nov 27	1	SCN	1.0	5200	1.5	18.0
18	Abell 09	05h 29m 00s	36° 02' 00"	Aur	Feb 2	Dec 3	2	SHN	0.50	13000	2.0	19.0
19	Abell 10	05h 31m 48s	06° 56' 09"	Ori	Feb 2	Dec 3	2	SRY/h	0.50	13000	2.0	14.0
26	Abell 12	06h 02m 23s	09° 39' 03"	Ori	Feb 10	Dec 11	2	SRN/i	0.70	7000	1.4	12.0
27	Abell 13	06h 04m 47s	03° 56' 27"	Ori	Feb 11	Dec 12	3	EHY/ir	3.0	4000	3.5	19.9
28	Abell 14	06h 11m 09s	11° 46' 47"	Ori	Feb 12	Dec 13	2	BHY/q	0.70	11000	2.1	15.0
31	Abell 15	06h 27m 02s	-25° 22' 54"	CMa	Feb 16	Dec 17	2	EOY/i	0.60	12000	2.0	16.0
34	Abell 16	06h 43m 55s	61° 47' 25"	Lyn	Feb 21	Dec 22	1	SCY/i	2.4	4400	3.0	18.7
35	Abell 18	06h 56m 14s	-02° 53' 08"	Mon	Feb 24	Dec 25	2	ERN/hr	1.4	5000	2.0	20.9
36	Abell 19	06h 59m 57s	14° 36' 47"	Gem	Feb 25	Dec 26	2	ERY/q	1.4	7500	3.0	16.0
42	Abell 20	07h 22m 58s	01° 45' 37"	CMi	Mar 2	Jan 1	1	SOY/r	1.0	6500	2.0	16.4
45	Abell 21	07h 29m 03s	13° 14' 48"	Gem	Mar 4	Jan 2	5	ARN/if	8.0	1500	3.5	16.0
48	Abell 22	07h 36m 06s	02° 24' 00"	CMi	Mar 6	Jan 4	3	BHN/c	2.0	2800	1.6	19.6
52	Abell 23	07h 43m 19s	-34° 45' 13"	Pup	Mar 8	Jan 6	1	SRN/ho	1.0	7500	2.2	u
55	Abell 24	07h 51m 38s	03° 00' 27"	CMi	Mar 10	Jan 8	3	AHY/q	7.0	2300	4.5	17.2
57	Abell 25	08h 06m 45s	-02° 52' 43"	Mon	Mar 14	Jan 12	2	EOY/q	3.0	2500	2.2	18.0
59	Abell 26	08h 09m 01s	-32° 40' 15"	Pup	Mar 14	Jan 12	2	SRN/hr	0.70	11000	2.1	18.0
60	Abell 27	08h 31m 53s	-32° 06' 07"	Pyx	Mar 20	Jan 18	2	EHN/hq	1.0	7700	2.1	16.0
63	Abell 28	08h 41m 35s	58° 13' 54"	UMa	Mar 22	Jan 21	1	AHY/i	8.0	1250	3.0	16.6
62	Abell 29	08h 40m 14s	-20° 53' 41"	Pyx	Mar 22	Jan 20	2	ARY/q	6.0	3600	6.0	18.3
64	Abell 30	08h 46m 54s	17° 52' 33"	Cnc	Mar 24	Jan 22	3	SOY/f	2.0	7000	4.0	14.3
65	Abell 31	08h 54m 13s	08° 53' 59"	Cnc	Mar 26	Jan 24	4	ARY/o	15	1800	8.0	15.5
68	Abell 33	09h 39m 09s	-02° 48' 33"	Hya	Apr 6	Feb 4	4	SOY/ot	4.0	2500	3.0	13.0
69	Abell 34	09h 45m 35s	-13° 10' 14"	Hya	Apr 8	Feb 6	2	SOY/qt	5.0	2400	3.3	16.3
79	Abell 36	13h 40m 41s	-19° 52' 57"	Vir	Jun 6	Apr 6	2	ACY/f	8.0	800	2.0	12.0
80	Abell 37	14h 04m 26s	-17° 13' 41"	Vir	Jun 12	Apr 12	2	ERY/h	1.0	8000	2.3	13.9
87	Abell 38	16h 23m 17s	-31° 44' 57"	Sco	Jul 17	May 18	3	EHN/f	2.5	2000	1.5	20.0
88	Abell 39	16h 27m 33s	27° 54' 34"	Her	Jul 19	May 19	4	SOY/ot	3.0	4000	3.5	15.6
90	Abell 40	16h 48m 34s	-21° 00' 40"	Oph	Jul 24	May 24	2	EOY/r	0.60	13000	2.2	18.0
96	Abell 41	17h 29m 04s	-15° 13' 21"	Ser	Aug 3	Jun 3	2	ERY/q	0.40	15000	1.7	16.0
98	Abell 42	17h 31m 31s	-08° 19' 10"	Oph	Aug 4	Jun 4	1	ECY/r	1.0	11000	3.0	20.0
102	Abell 43	17h 53m 32s	10° 37' 20"	Oph	Aug 9	Jun 10	3	SOY/ft	1.3	5000	2.0	15.0
111	Abell 44	18h 30m 11s	-16° 45' 27"	Sgr	Aug 19	Jun 19	2	BRN/q	1.0	7000	2.0	16.0
112	Abell 45	18h 30m 17s	-11° 36' 54"	Sct	Aug 19	Jun 19	2	PHN/f	7.0	u	u	13.0
113	Abell 46	18h 31m 19s	26° 56' 17"	Lyr	Aug 19	Jun 19	2	EOY/o	1.3	7200	2.7	15.0
116	Abell 47	18h 35m 22s	-00° 13' 32"	Ser	Aug 20	Jun 20	2	EHY/q	0.30	30000	2.6	19.0
118	Abell 48	18h 42m 49s	-03° 13' 00"	Aql	Aug 22	Jun 22	2	PHN	0.80	11800	2.7	17.0
120	Abell 49	18h 53m 29s	-06° 29' 14"	Sct	Aug 24	Jun 25	2	ERN/r	1.0	u	u	16.0
124	Abell 50	18h 59m 20s	48° 27' 57"	Dra	Aug 26	Jun 26	2	SOY/o	0.50	17000	2.4	16.6
125	Abell 51	19h 01m 01s	-18° 12' 16"	Sgr	Aug 26	Jun 27	2	SOY/r	1.0	6000	1.7	15.4
130	Abell 52	19h 04m 32s	17° 57' 10"	Aql	Aug 27	Jun 28	2	EOY/iq	1.0	6000	1.7	14.0
132	Abell 53	19h 06m 46s	06° 23' 56"	Aql	Aug 28	Jun 28	2	SRN/r	0.50	6300	1.0	14.0
133	Abell 54	19h 08m 39s	22° 58' 51"	Vul	Aug 28	Jun 29	1	ERY	1.0	9000	2.6	17.0

Alphabetic List of Top 250 PNe (2/5)

#	PN Name	RA (H/M/S)	DEC (D/M/S)	Const.	Transit 9:00 PM	Transit 1:00 AM	Score	Class	Size (')	Distance (ly)	Diam. (ly)	Visual Mag
134	Abell 55	19h 10m 30s	-02° 21' 02"	Aql	Aug 29	Jun 29	2	BRN/q	1.3	8200	3.0	13.0
136	Abell 56	19h 13m 07s	02° 52' 49"	Aql	Aug 29	Jun 30	2	EHY/hr	3.0	4900	4.0	u
139	Abell 57	19h 17m 04s	25° 37' 26"	Vul	Aug 30	Jul 1	2	EOY	1.0	7000	2.0	17.7
140	Abell 58	19h 18m 20s	01° 46' 51"	Aql	Aug 31	Jul 1	2	EHY/hi	0.80	u	u	u
143	Abell 59	19h 18m 41s	19° 33' 56"	Sge	Aug 31	Jul 1	2	EHN/i	1.8	4600	2.4	16.0
145	Abell 60	19h 19m 17s	-12° 14' 52"	Sgr	Aug 31	Jul 1	2	EOY	1.7	5600	2.7	16.0
144	Abell 61	19h 19m 10s	46° 14' 36"	Cyg	Aug 31	Jul 1	2	SOY/t	3.0	3000	2.6	17.4
150	Abell 62	19h 33m 18s	10° 37' 01"	Aql	Sep 4	Jul 5	3	ERY/r	3.0	1600	1.4	15.0
154	Abell 63	19h 42m 10s	17° 05' 08"	Sge	Sep 6	Jul 7	2	PCY	0.70	9000	1.8	17.0
158	Abell 65	19h 46m 34s	-23° 08' 12"	Sgr	Sep 7	Jul 8	3	EOY	4.0	4000	5.0	15.8
162	Abell 66	19h 57m 32s	-21° 36' 37"	Sgr	Sep 10	Jul 11	2	ERY	4.5	2000	2.5	17.4
164	Abell 67	19h 58m 29s	03° 02' 23"	Aql	Sep 10	Jul 11	2	EON/q	1.0	6500	2.0	18.0
167	Abell 68	20h 00m 11s	21° 42' 58"	Vul	Sep 10	Jul 12	2	ECY/q	0.70	6800	1.5	15.0
176	Abell 69	20h 19m 56s	38° 24' 31"	Cyg	Sep 15	Jul 17	2	EHN/r	0.40	14000	1.7	u
178	Abell 70	20h 31m 33s	-07° 05' 21"	Aql	Sep 18	Jul 20	2	ERY/hr	1.0	12000	3.5	15.0
180	Abell 71	20h 32m 23s	47° 21' 04"	Cyg	Sep 19	Jul 20	2	PHY/h	3.0	2400	2.0	19.3
183	Abell 72	20h 50m 02s	13° 33' 28"	Del	Sep 23	Jul 24	3	BOY/f	3.0	3700	3.0	16.0
185	Abell 73	20h 56m 26s	57° 25' 56"	Cep	Sep 25	Jul 26	2	ERY/q	1.4	6000	2.4	17.0
193	Abell 74	21h 16m 52s	24° 08' 51"	Vul	Sep 30	Jul 31	2	ARY/r	9.0	2200	5.7	17.1
196	Abell 75	21h 26m 24s	62° 53' 28"	Cep	Oct 2	Aug 2	2	EOY	1.0	6000	1.7	15.0
198	Abell 78	21h 35m 29s	31° 41' 45"	Cyg	Oct 5	Aug 5	3	PRY/f	2.0	2300	1.3	13.0
209	Abell 79	22h 26m 17s	54° 49' 41"	Lac	Oct 17	Aug 18	3	PHY/f	2.0	4000	2.3	17.0
211	Abell 80	22h 34m 46s	52° 26' 04"	Lac	Oct 20	Aug 20	2	EHY/q	2.2	6000	3.8	16.0
213	Abell 81	22h 42m 25s	80° 26' 33"	Cep	Oct 22	Aug 22	2	SOY/r	0.70	14000	2.8	14.0
219	Abell 82	23h 45m 47s	57° 04' 01"	Cas	Nov 7	Sep 7	2	ERN/q	1.7	6500	3.2	13.0
220	Abell 83	23h 46m 46s	54° 44' 40"	Cas	Nov 7	Sep 7	1	ERN/h	0.70	u	u	15.0
222	Abell 84	23h 47m 45s	51° 23' 58"	Cas	Nov 7	Sep 7	2	ERY/hq	2.7	5000	4.0	18.6
223	Abell 86	00h 01m 33s	70° 42' 42"	Cep	Nov 11	Sep 11	2	SHN/i	1.3	7300	3.0	u
6	Baade 1	03h 53m 37s	19° 29' 39"	Tau	Jan 8	Nov 9	2	SOY/r	1.0	8100	2.4	14.3
227	Böhm-Vitense 5-1	00h 19m 59s	62° 59' 06"	Cas	Nov 15	Sep 15	2	PCN/fi	1.5	7200	3.0	u
230	Böhm-Vitense 5-2	00h 40m 24s	62° 51' 00"	Cas	Nov 20	Sep 21	3	PRN/i	0.70	650	0.14	21.5
236	Böhm-Vitense 5-3	01h 53m 03s	56° 24' 00"	Per	Dec 9	Oct 9	2	ERN/r	0.50	23000	3.5	15.0
151	Campbell's Star	19h 34m 45s	30° 30' 59"	Cyg	Sep 4	Jul 5	2	XHY	0.20	10000	0.65	10.5
23	DeHt 1	05h 55m 00s	-22° 30' 00"	Lep	Feb 8	Dec 9	2	SOY/it	2.0	6200	3.5	u
148	DeHt 4	19h 26m 27s	13° 19' 35"	Aql	Sep 2	Jul 3	2	PHN/i	1.5	u	u	u
206	DeHt 5	22h 19m 34s	70° 56' 01"	Cep	Oct 16	Aug 16	2	ARY/i	9.0	1000	2.8	15.5
184	Ear Nebula	20h 50m 13s	46° 55' 00"	Cyg	Sep 23	Jul 24	3	PRN/fi	6.0	u	u	u
233	EGB 01	01h 07m 06s	73° 33' 00"	Cas	Nov 27	Sep 27	3	ARY/i	3.5	1000	1.0	16.4
70	EGB 06	09h 53m 00s	13° 45' 00"	Leo	Apr 10	Feb 8	1	ARY/iq	12	2100	8.0	16.0
138	ETHOS 1	19h 16m 30s	36° 08' 56"	Lyr	Aug 30	Jul 1	3	BOY/fj	1.0	u	u	u
237	Ferrero 6	01h 56m 26s	65° 28' 20"	Cas	Dec 10	Oct 10	2	ERY	3.3	u	u	u
201	G100.4+04.6	21h 40m 00s	58° 59' 00"	Cep	Oct 6	Aug 6	2	PRY/y	1.0	u	u	u
242	G132.8+2.0	02h 20m 45s	63° 11' 34"	Cas	Dec 16	Oct 16	2	ERN	0.50	4300	0.70	u
12	G156.4+01.1	04h 38m 25s	48° 38' 52"	Per	Jan 20	Nov 20	1	PHN/i	2	u	u	u
95	Haro 2-08	17h 24m 46s	-21° 33' 36"	Oph	Aug 2	Jun 2	2	BCN	0.20	u	u	u
11	Haro 3-29	04h 37m 24s	25° 02' 44"	Tau	Jan 20	Nov 20	2	PCY	0.30	15000	1.4	15.9
20	Haro 3-75	05h 40m 42s	12° 21' 00"	Ori	Feb 5	Dec 6	2	ERY	0.50	10000	1.5	14.0
128	HaTr 11	19h 02m 59s	03° 02' 21"	Aql	Aug 27	Jun 27	2	MRN	1.0	u	u	u
1	HDW 03	03h 27m 15s	45° 24' 20"	Per	Jan 2	Nov 2	3	ACN	9.0	3000	8.0	17.2
30	HDW 05	06h 23m 36s	-10° 13' 00"	Mon	Feb 15	Dec 17	2	MHN/i	1.5	3600	1.5	15.4

Alphabetic List of Top 250 PNe (3/5)

#	PN Name	RA (H / M / S)	DEC (D / M / S)	Const.	Transit 9:00 PM	Transit 1:00 AM	Score	Class	Size (')	Distance (ly)	Diam. (ly)	Visual Mag
163	HDW 12	19h 58m 12s	-26° 28' 16"	Sgr	Sep 10	Jul 11	2	SHY	0.80	2100	0.50	18.5
165	Henize 1-4	19h 59m 18s	31° 33' 00"	Cyg	Sep 10	Jul 11	1	BHN	0.40	11000	1.4	u
248	HFG 1	03h 01m 00s	64° 58' 00"	Cas	Dec 26	Oct 26	3	AOY	8.0	2300	5.0	14.6
197	Humason 1-2	21h 33m 06s	39° 38' 00"	Cyg	Oct 4	Aug 4	3	PON/y	0.50	9000	1.4	12.0
249	IC 0289	03h 10m 19s	61° 19' 01"	Cas	Dec 28	Oct 29	2	BCY	0.75	5000	1.1	16.0
4	IC 0351	03h 47m 33s	35° 02' 48"	Per	Jan 7	Nov 7	1	XON	0.20	18000	1.0	11.9
17	IC 0418	05h 27m 28s	-12° 41' 50"	Lep	Feb 1	Dec 2	1	ERY	0.20	5000	0.30	9.3
122	IC 1295	18h 54m 37s	-08° 49' 36"	Sct	Aug 25	Jun 25	3	SOY/ot	2.0	3400	2.0	12.5
238	IC 1747	01h 57m 36s	63° 19' 18"	Cas	Dec 10	Oct 10	2	EOY	0.20	10000	0.60	15.4
7	IC 2003	03h 56m 22s	33° 52' 30"	Per	Jan 9	Nov 9	1	XON	0.1	18000	0.6	12.0
24	IC 2149	05h 56m 24s	46° 06' 17"	Aur	Feb 9	Dec 10	2	BOY/y	0.25	3600	0.28	11.4
77	IC 3568	12h 33m 07s	82° 33' 49"	Cam	May 20	Mar 20	1	EOY	0.20	12000	0.70	10.6
81	IC 4406	14h 22m 26s	-44° 09' 04"	Lup	Jun 17	Apr 17	4	BRN	1.8	9000	5.0	11.0
85	IC 4593	16h 11m 44s	12° 04' 19"	Her	Jul 15	May 15	1	XOY	0.10	10500	0.30	10.7
91	IC 4634	17h 01m 34s	-21° 49' 30"	Oph	Jul 27	May 27	2	BOY/y	0.20	13000	0.80	11.3
203	IC 5148	21h 59m 35s	-39° 23' 09"	Gru	Oct 11	Aug 11	4	ECY/r	2.0	3000	1.8	16.5
208	IC 5217	22h 23m 56s	50° 58' 00"	Lac	Oct 17	Aug 17	1	XON	0.10	18000	0.6	11.3
5	IsWe 1	03h 49m 00s	50° 00' 00"	Per	Jan 7	Nov 7	2	ACN/i	20	1400	8.0	16.5
204	IsWe 2	22h 13m 00s	65° 54' 00"	Cep	Oct 14	Aug 14	1	AHY	20	850	5.0	18.2
82	Jacoby 1	15h 22m 00s	52° 22' 00"	Boo	Jul 2	May 2	1	AOY	10	2600	7.0	16.6
217	Jones 1	23h 35m 53s	30° 28' 06"	Peg	Nov 4	Sep 4	3	EOY/iq	5.0	2600	4.0	15.6
56	Jones-Emberson 1	07h 57m 54s	53° 25' 00"	Lyn	Mar 11	Jan 10	4	ERY/qr	6.0	1600	3.0	17.0
174	Ju 1	20h 15m 26s	38° 02' 48"	Cyg	Sep 14	Jul 15	3	SON/t	4.0	4000	5.0	u
214	KjPn 8	23h 24m 00s	60° 57' 00"	Cas	Nov 1	Sep 1	4	MHY/fy	11	6000	20	u
66	Kohoutek 1-02	08h 57m 48s	-28° 36' 00"	Pyx	Mar 27	Jan 25	3	POY/jy	1.8	9000	5.0	15.3
39	Kohoutek 1-10	07h 12m 36s	-16° 06' 00"	CMa	Feb 28	Dec 29	2	BHN	1.5	9000	4.0	u
54	Kohoutek 1-12	07h 50m 12s	-19° 18' 16"	Pup	Mar 9	Jan 8	2	ERN/q	1.0	9000	2.7	16.0
99	Kohoutek 1-14	17h 42m 36s	21° 30' 00"	Her	Aug 7	Jun 7	2	SOY/t	1.0	11000	3.0	u
129	Kohoutek 1-17	19h 03m 36s	19° 12' 00"	Sge	Aug 27	Jun 27	3	BOY/r	0.80	10000	2.2	u
218	Kohoutek 1-20	23h 39m 10s	48° 12' 31"	And	Nov 5	Sep 5	2	EON/r	0.70	14000	2.7	16.5
75	Kohoutek 1-22	11h 26m 42s	-34° 22' 00"	Hya	May 3	Mar 3	3	SOY/ot	3.0	3000	2.8	16.1
73	Kohoutek 1-28	10h 34m 30s	-29° 07' 00"	Hya	Apr 20	Feb 18	1	SOY	1.0	6600	2.0	16.0
16	Kohoutek 2-01	05h 07m 16s	30° 48' 00"	Aur	Jan 27	Nov 27	2	EOY/i	3.0	3500	3.0	u
152	Kohoutek 2-07	19h 41m 25s	-20° 24' 44"	Sgr	Sep 6	Jul 7	1	AOY	2.5	6000	4.4	u
159	Kohoutek 3-46	19h 50m 00s	33° 28' 00"	Cyg	Sep 8	Jul 9	2	BHN/c	1.0	u	u	18.5
239	Kohoutek 3-91	01h 58m 36s	66° 34' 00"	Cas	Dec 10	Oct 10	1	XHN	0.30	28000	2.5	21.0
240	Kohoutek 3-92	02h 03m 40s	64° 57' 36"	Cas	Dec 12	Oct 12	2	BRN/c	0.30	22000	2.0	17.0
243	Kohoutek 3-93	02h 26m 30s	65° 47' 53"	Cas	Dec 17	Oct 18	1	XRN	0.20	u	u	18.0
3	Kohoutek 3-94	03h 36m 08s	60° 03' 47"	Cam	Jan 4	Nov 4	1	BRN	0.20	22000	1.4	16.0
181	Kohoutek 4-53	20h 42m 18s	37° 24' 00"	Cyg	Sep 21	Jul 22	1	PHN	0.30	u	u	u
182	Kohoutek 4-55	20h 45m 12s	44° 24' 00"	Cyg	Sep 22	Jul 23	3	BRN	0.50	4500	0.70	16.5
147	Kronberger 61	19h 21m 39s	38° 18' 57"	Lyr	Sep 1	Jul 2	2	SOY/t	5.0	13000	18	u
47	KW 8	07h 33m 25s	-23° 26' 09"	Pup	Mar 5	Jan 3	1	EHY/i	2.0	u	u	u
78	LoTr 5	12h 56m 00s	25° 53' 00"	Com	May 26	Mar 26	2	AOY/o	9.0	1600	4.2	8.9
166	M 027	19h 59m 36s	22° 43' 15"	Vul	Sep 10	Jul 11	5	BRY/c	15	1200	5.0	14.1
121	M 057	18h 53m 35s	33° 01' 44"	Lyr	Aug 25	Jun 25	5	BRY/r	1.5	2300	1.0	15.8
235	M 076	01h 42m 18s	51° 34' 15"	Per	Dec 6	Oct 6	4	BRN	3.0	3000	2.7	17.5
74	M 097	11h 14m 48s	55° 01' 08"	UMa	Apr 30	Feb 28	5	BRY/o	3.8	2000	1.8	15.8
33	Minkowski 1-07	06h 37m 21s	24° 00' 36"	Gem	Feb 19	Dec 20	2	BRN	0.50	20000	3.0	13.5
51	Minkowski 1-18	07h 42m 06s	-14° 12' 00"	Pup	Mar 7	Jan 6	2	ERN/ir	0.50	15000	2.2	14.0

Appendix A

Alphabetic List of Top 250 PNe (4/5)

#	PN Name	RA (H / M / S)	DEC (D / M / S)	Const.	Transit 9:00 PM	Transit 1:00 AM	Score	Class	Size (')	Distance (ly)	Diam. (ly)	Visual Mag
100	Minkowski 1-28	17h 47m 36s	-22° 04' 00"	Sgr	Aug 8	Jun 8	3	BHN/c	0.90	15000	4.0	17.0
105	Minkowski 1-41	18h 09m 30s	-24° 12' 15"	Sgr	Aug 13	Jun 14	3	PHN/c	2.0	2800	1.8	15.0
117	Minkowski 1-57	18h 40m 20s	-10° 39' 47"	Sct	Aug 21	Jun 21	2	BRN/c	0.50	15000	2.3	14.0
119	Minkowski 1-64	18h 50m 00s	35° 15' 00"	Lyr	Aug 24	Jun 24	2	EHN/r	0.40	18000	2.3	u
170	Minkowski 1-75	20h 04m 42s	31° 15' 00"	Cyg	Sep 12	Jul 13	3	MRN/c	0.30	14000	1.2	u
200	Minkowski 1-79	21h 37m 02s	48° 56' 06"	Cyg	Oct 5	Aug 5	3	MRN/c	0.80	8600	2.0	19.0
160	Minkowski 2-48	19h 50m 28s	25° 54' 22"	Vul	Sep 8	Jul 9	3	BHN/c	0.50	21000	3.0	u
205	Minkowski 2-51	22h 16m 03s	57° 28' 31"	Cep	Oct 15	Aug 15	2	BRN/r	1.0	6000	1.8	u
207	Minkowski 2-52	22h 20m 31s	57° 36' 18"	Cep	Oct 16	Aug 16	3	MRN	0.50	15000	2.2	14.0
216	Minkowski 2-55	23h 31m 54s	70° 22' 00"	Cep	Nov 3	Sep 3	3	BRN/o	1.0	7800	2.2	u
37	Minkowski 3-01	07h 02m 50s	-31° 35' 30"	CMa	Feb 25	Dec 27	3	BON/y	0.50	15000	2.2	13.0
44	Minkowski 3-03	07h 26m 34s	-05° 21' 52"	Mon	Mar 3	Jan 2	2	BRN	0.30	19000	1.6	u
114	Minkowski 3-28	18h 32m 41s	-10° 05' 48"	Sct	Aug 19	Jun 19	3	BRN/c	0.40	16000	2.0	15.0
115	Minkowski 3-55	18h 33m 15s	-10° 15' 07"	Sct	Aug 19	Jun 20	1	BRN/c	0.20	u	u	20.0
108	Minkowski 4-09	18h 14m 18s	-04° 59' 22"	Ser	Aug 15	Jun 15	2	ERY	0.70	6000	1.3	16.6
146	Minkowski 4-14	19h 21m 01s	07° 36' 59"	Aql	Aug 31	Jul 2	1	BHN	0.20	25000	1.6	u
171	Minkowski 4-17	20h 09m 00s	43° 44' 00"	Cyg	Sep 13	Jul 14	2	BRN/cr	1.0	u	u	u
195	MWP 1	21h 17m 08s	34° 12' 27"	Cyg	Sep 30	Aug 1	2	AOY/c	13	4500	17	12.3
156	Necklace Nebula	19h 43m 59s	17° 09' 02"	Sge	Sep 6	Jul 8	2	POY	0.50	15000	2.2	11.0
226	NGC 0040	00h 13m 01s	72° 31' 19"	Cep	Nov 13	Sep 14	4	BHY/cf	0.60	3500	0.60	11.5
232	NGC 0246	00h 47m 03s	-11° 52' 19"	Cet	Nov 22	Sep 22	4	BOY/o	4.5	1600	2.2	11.8
2	NGC 1360	03h 33m 15s	-25° 52' 19"	For	Jan 3	Nov 3	4	POY/y	9.0	1200	3.0	11.0
8	NGC 1501	04h 06m 59s	60° 55' 14"	Cam	Jan 12	Nov 12	4	EOY/f	1.0	5700	1.7	13.0
9	NGC 1514	04h 09m 17s	30° 46' 33"	Tau	Jan 12	Nov 13	4	SOY/o	3.0	2000	1.9	9.5
10	NGC 1535	04h 14m 16s	-12° 44' 22"	Eri	Jan 14	Nov 14	3	BOY	0.80	4000	1.0	12.8
21	NGC 2022	05h 42m 06s	09° 05' 13"	Ori	Feb 5	Dec 6	2	BOY	0.50	8000	1.1	14.2
32	NGC 2242	06h 34m 07s	44° 46' 38"	Aur	Feb 18	Dec 19	1	BOY/o	0.50	9600	1.5	15.0
38	NGC 2346	07h 09m 22s	-00° 48' 22"	Mon	Feb 27	Dec 28	3	BRY/c	2.0	4800	3.0	11.6
43	NGC 2371	07h 25m 34s	29° 29' 17"	Gem	Mar 3	Jan 1	4	BOY/cy	2.0	5000	3.2	13.5
46	NGC 2392	07h 29m 11s	20° 54' 42"	Gem	Mar 4	Jan 2	4	BHY/f	1.0	6000	1.7	9.7
49	NGC 2438	07h 41m 51s	-14° 44' 05"	Pup	Mar 7	Jan 5	3	ERY	1.5	4200	1.9	11.5
50	NGC 2440	07h 41m 55s	-18° 12' 31"	Pup	Mar 7	Jan 5	3	MRN	1.3	4000	1.6	10.1
53	NGC 2452	07h 47m 26s	-27° 20' 07"	Pup	Mar 9	Jan 7	2	PON	0.50	11400	1.7	12.0
61	NGC 2610	08h 33m 23s	-16° 08' 57"	Hya	Mar 20	Jan 19	2	BOY/r	1.2	7000	2.5	12.7
67	NGC 2818	09h 16m 01s	-36° 37' 37"	Pyx	Mar 31	Jan 29	4	BRN	2.0	10000	6.0	11.2
71	NGC 3132	10h 07m 02s	-40° 26' 11"	Vel	Apr 13	Feb 11	4	MRY/r	1.2	2800	1.0	10.0
72	NGC 3242	10h 24m 46s	-18° 38' 34"	Hya	Apr 18	Feb 16	4	EOY/r	1.0	4000	1.1	12.1
76	NGC 4361	12h 24m 31s	-18° 47' 06"	Crv	May 18	Mar 18	3	BOY/j	1.8	1250	0.70	13.2
83	NGC 6026	16h 01m 21s	-34° 32' 37"	Lup	Jul 12	May 12	2	EOY/hr	1.0	9500	3.0	12.5
84	NGC 6058	16h 04m 27s	40° 40' 59"	Her	Jul 13	May 13	2	MOY/r	0.70	11500	2.2	12.9
86	NGC 6072	16h 12m 58s	-36° 13' 47"	Sco	Jul 15	May 15	2	BRY	1.8	4000	2.0	12.1
89	NGC 6210	16h 44m 30s	23° 47' 59"	Her	Jul 23	May 23	2	BOY/fjy	0.80	6500	1.4	11.7
92	NGC 6302	17h 13m 44s	-37° 06' 12"	Sco	Jul 30	May 30	4	MRN	2.5	4000	3.0	10.1
93	NGC 6309	17h 14m 04s	-12° 54' 38"	Oph	Jul 30	May 31	2	BON/j	0.60	9000	1.5	11.5
94	NGC 6337	17h 22m 16s	-38° 29' 00"	Sco	Aug 1	Jun 2	4	BRY	1.0	5000	1.4	12.3
97	NGC 6369	17h 29m 21s	-23° 45' 34"	Oph	Aug 3	Jun 3	1	BRY	1.2	3000	1.0	12.0
101	NGC 6445	17h 49m 15s	-20° 00' 36"	Sgr	Aug 8	Jun 8	3	MRN	2.5	4500	3.4	18.7
104	NGC 6537	18h 05m 13s	-19° 50' 35"	Sgr	Aug 12	Jun 12	2	BHY/cf	2.0	5000	3.0	13.6
103	NGC 6543	17h 58m 33s	66° 37' 59"	Dra	Aug 11	Jun 11	5	ERN	5.0	3300	5.0	11.3

Alphabetic List of Top 250 PNe (5/5)

#	PN Name	RA (H / M / S)	DEC (D / M / S)	Const.	Transit 9:00 PM	Transit 1:00 AM	Score	Class	Size (')	Distance (ly)	Diam. (ly)	Visual Mag
107	NGC 6572	18h 12m 06s	06° 51' 13"	Oph	Aug 14	Jun 14	1	EON	0.25	5000	0.40	10.8
110	NGC 6629	18h 25m 42s	-23° 12' 10"	Sgr	Aug 17	Jun 18	2	BOY/r	0.40	7700	1.0	10.0
127	NGC 6741	19h 02m 37s	-00° 26' 57"	Aql	Aug 27	Jun 27	1	ECN	0.15	12000	0.50	11.5
131	NGC 6751	19h 05m 56s	-05° 59' 31"	Aql	Aug 28	Jun 28	3	BRY/f	0.70	8400	1.8	15.5
135	NGC 6765	19h 11m 07s	30° 32' 45"	Lyr	Aug 29	Jun 29	2	PON	1.0	5000	1.5	12.9
137	NGC 6772	19h 14m 36s	-02° 42' 24"	Aql	Aug 30	Jun 30	2	ERY	1.5	4000	1.7	16.8
141	NGC 6778	19h 18m 25s	-01° 35' 48"	Aql	Aug 31	Jul 1	2	BHY/c	0.50	10000	1.5	14.8
142	NGC 6781	19h 18m 28s	06° 32' 20"	Aql	Aug 31	Jul 1	4	BHY	1.9	3000	1.8	11.8
149	NGC 6804	19h 31m 35s	09° 13' 31"	Aql	Sep 3	Jul 4	2	BOY/r	2.5	2300	1.8	13.4
155	NGC 6818	19h 43m 58s	-14° 09' 10"	Sgr	Sep 6	Jul 7	2	BOY/r	0.40	5000	0.60	9.3
157	NGC 6826	19h 44m 48s	50° 31' 30"	Cyg	Sep 7	Jul 8	4	BOY/r	0.50	4200	0.65	9.6
161	NGC 6842	19h 55m 02s	29° 17' 20"	Vul	Sep 9	Jul 10	2	EOY	1.0	4500	1.4	16.0
168	NGC 6852	20h 00m 39s	01° 43' 41"	Aql	Sep 11	Jul 12	2	BOY	0.50	9000	1.4	17.5
173	NGC 6891	20h 15m 09s	12° 42' 16"	Del	Sep 14	Jul 15	2	EOY	0.30	12000	1.1	12.5
175	NGC 6894	20h 16m 24s	30° 33' 55"	Cyg	Sep 15	Jul 16	2	BRY	0.70	5000	1.0	12.3
177	NGC 6905	20h 23m 23s	20° 06' 16"	Del	Sep 16	Jul 17	2	BOY/fy	1.5	5750	2.5	14.5
186	NGC 7008	21h 00m 33s	54° 32' 35"	Cyg	Sep 26	Jul 27	4	PBY	1.8	3000	1.5	12.0
187	NGC 7009	21h 04m 11s	-11° 21' 50"	Aql	Sep 27	Jul 28	4	BOY/y	0.70	4500	1.0	12.0
188	NGC 7026	21h 06m 19s	47° 51' 08"	Cyg	Sep 27	Jul 28	4	MRY/f	0.70	10000	2.0	15.0
189	NGC 7027	21h 07m 02s	42° 14' 10"	Cyg	Sep 27	Jul 29	2	PBN	0.25	3000	0.22	8.8
192	NGC 7048	21h 14m 15s	46° 17' 18"	Cyg	Sep 29	Jul 30	4	BRY/c	1.0	5000	1.5	12.1
199	NGC 7094	21h 36m 53s	12° 47' 19"	Peg	Oct 5	Aug 5	3	EOY/f	1.8	5400	2.8	13.7
202	NGC 7139	21h 46m 09s	63° 47' 30"	Cep	Oct 7	Aug 7	2	ERN/h	1.0	4000	1.3	13.3
210	NGC 7293	22h 29m 38s	-20° 50' 13"	Aql	Oct 18	Aug 18	5	BRY/r	12	650	2.5	13.5
212	NGC 7354	22h 40m 20s	61° 17' 07"	Cep	Oct 21	Aug 21	3	EOY	0.40	4200	0.50	12.2
215	NGC 7662	23h 25m 54s	42° 32' 06"	And	Nov 2	Sep 2	3	EHY/y	0.50	6400	1.0	8.3
228	OU 2	00h 30m 57s	61° 24' 34"	Cas	Nov 18	Sep 18	2	EOY/r	1.2	u	u	u
153	PC 22	19h 42m 04s	13° 50' 35"	Aql	Sep 6	Jul 7	2	MON/c	1.0	17000	5.0	u
41	PFP 1	07h 22m 00s	-06° 18' 00"	Mon	Mar 2	Dec 31	1	ARY	25	1700	12	u
22	Pu 1	05h 52m 48s	28° 06' 00"	Tau	Feb 8	Dec 9	2	EHN	1.0	6800	2.0	18.0
58	Sa 2-21	08h 08m 42s	-19° 08' 00"	Pup	Mar 14	Jan 12	2	ERN/o	0.90	11000	3.0	13.7
191	Sh1-089	21h 14m 06s	47° 46' 00"	Cyg	Sep 29	Jul 30	3	BRN/c	3.0	6400	6.0	u
224	Sh1-118	00h 07m 20s	64° 57' 21"	Cas	Nov 12	Sep 12	1	EHN	3.0	4000	3.4	13.9
109	Sh2-068	18h 24m 58s	00° 51' 36"	Ser	Aug 17	Jun 17	3	ACY/i	8.0	1300	3.3	16.6
126	Sh2-071	19h 02m 00s	02° 09' 11"	Aql	Aug 27	Jun 27	3	MHY	3.0	3000	2.5	14.0
221	Sh2-174	23h 47m 08s	80° 49' 22"	Cep	Nov 7	Sep 7	4	ACY/i	17	1000	5.0	14.7
234	Sh2-188	01h 30m 33s	58° 24' 51"	Cas	Dec 3	Oct 3	4	ARY/fi	9.0	850	2.1	17.4
250	Sh2-200	03h 11m 01s	62° 47' 45"	Cas	Dec 29	Oct 29	4	AON/io	6.0	3600	6.0	u
13	Sh2-216	04h 43m 21s	46° 42' 06"	Per	Jan 21	Nov 21	3	ARN/i	120	420	15	12.9
244	WeBo 1	02h 40m 14s	61° 09' 17"	Cas	Dec 21	Oct 21	2	EHY	1.3	5120	2.0	14.5
25	WeDe 1	05h 59m 00s	10° 42' 00"	Ori	Feb 9	Dec 10	2	AHY	15	1900	8.0	17.2
229	Weinberger 1-01	00h 38m 54s	66° 23' 49"	Cas	Nov 20	Sep 20	1	EHN	0.30	19000	1.6	u
29	Weinberger 1-04	06h 14m 34s	07° 34' 30"	Ori	Feb 13	Dec 14	2	BHN/c	0.70	15000	3.0	u
40	Weinberger 1-06	07h 17m 24s	-10° 07' 00"	Mon	Mar 1	Dec 30	1	EHY	1.5	4600	2.0	u
172	Weinberger 1-09	20h 09m 05s	26° 26' 56"	Vul	Sep 13	Jul 14	1	ERN	0.40	u	u	u
179	Weinberger 1-10	20h 31m 54s	48° 53' 00"	Cyg	Sep 18	Jul 20	2	SBN/t	3.0	u	u	u
190	Weinberger 1-11	21h 10m 54s	50° 30' 00"	Cyg	Sep 28	Jul 30	1	EHN/r	0.40	u	u	u
194	Weinberger 2-245	21h 18m 06s	43° 30' 00"	Cyg	Sep 30	Jul 31	1	PHN	1.2	u	u	u
169	WeSb 5	20h 02m 42s	19° 36' 00"	Sge	Sep 11	Jul 12	2	SBN	2.5	u	u	17.6
123	YM 16	18h 54m 57s	06° 02' 31"	Ser	Aug 25	Jun 25	2	EHY/i	6.0	u	u	u

www.ingramcontent.com/pod-product-compliance
Lightning Source LLC
Chambersburg PA
CBHW051157220526
45473CB00003B/804